TWENTY FIRST CENTURY
science

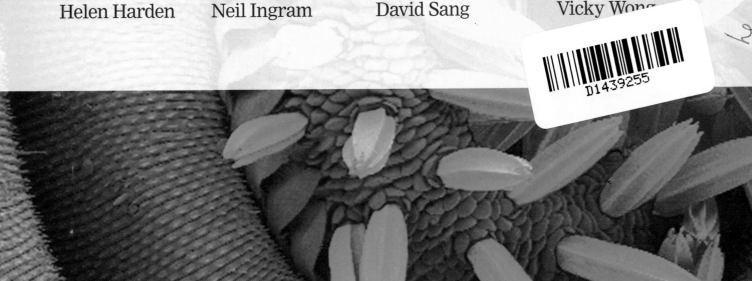

Project Directors

Angela Hall   Emma Palmer

Robin Millar   Mary Whitehouse

Editors

Emma Palmer   Carol Usher

Anne Scott   Mary Whitehouse

Authors

Ann Fullick   Andrew Hunt   Jacqueline Punter   Elizabeth Swinbank

Helen Harden   Neil Ingram   David Sang   Vicky Wong

THE UNIVERSITY *of* York

THE SALTERS' INSTITUTE

Nuffield Foundation

OCR
RECOGNISING ACHIEVEMENT

OXFORD
UNIVERSITY PRESS

Official Publisher Partnership

Contents

How to use this book

Welcome to Twenty First Century Science. This book has been specially written by a partnership between OCR, the University of York Science Education Group, the Nuffield Foundation Curriculum Programme, and Oxford University Press.

On these two pages you can see the types of page you will find in this book, and the features on them. Everything in the book is designed to provide you with the support you need to help you prepare for your examinations and achieve your best.

Module Openers

Why study?: This explains how what you're about to learn is relevant to everyday life.

Find out about: Every module starts with a short list of the things you'll be covering.

Ideas about Science: Here you can read the key ideas about science covered in this module.

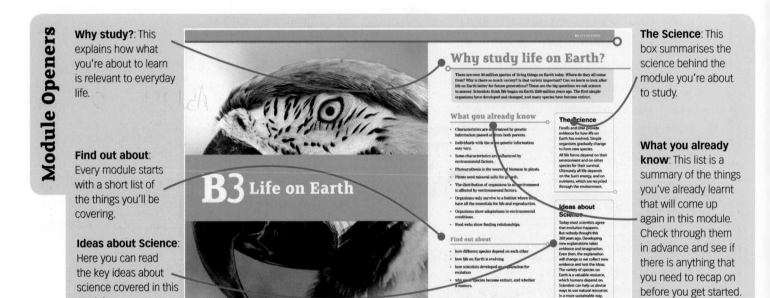

The Science: This box summarises the science behind the module you're about to study.

What you already know: This list is a summary of the things you've already learnt that will come up again in this module. Check through them in advance and see if there is anything that you need to recap on before you get started.

Main Pages

Find out about: For every part of the book you can see a list of the key points explored in that section.

Key words: The words in these boxes are the terms you need to understand for your exams. You can look for these words in the text in bold or check the glossary to see what they mean.

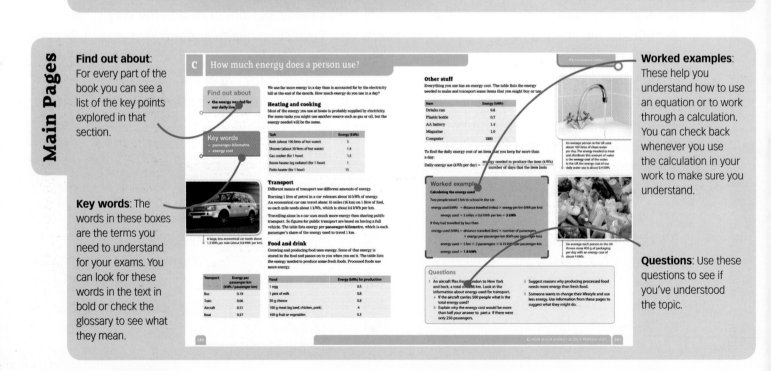

Worked examples: These help you understand how to use an equation or to work through a calculation. You can check back whenever you use the calculation in your work to make sure you understand.

Questions: Use these questions to see if you've understood the topic.

You should know: This is a summary of the main ideas in the unit. You can use it as a starting point for revision, to check that you know about the big ideas covered.

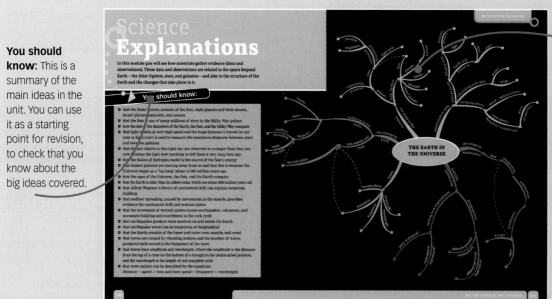

Visual summary: Another way to start revision is to use a visual summary, linking ideas together in groups so that you can see how one topic relates to another. You can use this page as a starting point for your own summary.

Ideas about Science: For every module this page summarises the ideas about science that you need to understand.

Review Questions: You can begin to prepare for your exams by using these questions to test how well you know the topics in this module.

Structure of assessment

Matching your course

What's in each module?

As you go through the book you should use the module opener pages to understand what you will be learning and why it is important. The table below gives an overview of the main topics each module includes.

B1
• What are genes and how do they affect the way that organisms develop?
• Why can people look like their parents, brothers and sisters, but not be identical to them?
• How can and should genetic information be used? How can we use our knowledge of genes to prevent disease?
• How is a clone made?

C1
• Which chemicals make up air, and which ones are pollutants? How do I make sense of data about air pollution?
• What chemical reactions produce air pollutants? What happens to these pollutants in the atmosphere?
• What choices can we make personally, locally, nationally or globally to improve air quality?

P1
• What do we know about the place of the Earth in the Universe?
• What do we know about the the Earth and how it is changing?

B2
• How do our bodies resist infection?
• What are vaccines and antibiotics and how do they work?
• What factors increase the risk of heart disease?
• How do our bodies keep a healthy water balance?

C2
• How do we measure the properties of materials and why are the results useful?
• Why is crude oil important as a source of new materials such as plastics and fibres?
• Why does it help to know about the molecular structure of materials such as plastics and fibres?
• What is nanotechnology and why is it important?

P2
• What types of electromagnetic radiation are there?
• Which types of electromagnetic radiation harm living tissue and why?
• What is the evidence for global warming, why might it be occuring? How serious a threat is it?
• How are electromagnetic waves used in communications?

B3
• Systems in balance – how do different species depend on each other?
• How has life on Earth evolved?
• What is the importance of biodiversity?

C3
• What were the origins of minerals in Britain that contribute to our economic wealth?
• Where does salt come from; why is it important?
• Why do we need chemicals such as alkalis and chlorine and how do we make them?
• What can we do to make our use of chemicals safe and sustainable?

P3
• How much energy do we use?
• How can electricity be generated?
• Which energy sources should we choose?

How do the modules fit together?

The modules in this book have been written to match the specification for GCSE Science. In the diagram to the right you can see that the modules can also be used to study parts of GCSE Biology, GCSE Chemistry, and GCSE Physics courses.

	GCSE Biology	GCSE Chemistry	GCSE Physics
GCSE Science	B1	C1	P1
	B2	C2	P2
	B3	C3	P3
GCSE Additional Science	B4	C4	P4
	B5	C5	P5
	B6	C6	P6
	B7	C7	P7

GCSE Science assessment

The content in the modules of this book matches the modules of the specification.

Twenty First Century Science offers two routes to the GCSE Science qualification, which includes different exam papers depending on the route you take.

The diagrams below show you which modules are included in each exam paper. They also show you how much of your final mark you will be working towards in each paper.

Unit	Modules Tested			Percentage	Type	Time	Marks Available
Route 1							
A161	B1	B2	B3	25%	Written Exam	1 h	60
A171	C1	C2	C3	25%	Written Exam	1 h	60
A181	P1	P2	P3	25%	Written Exam	1 h	60
A144	Controlled Assessment			25%		9 h	64

Unit	Modules Tested			Percentage	Type	Time	Marks Available
Route 2							
A141	B1	C1	P1	25%	Written Exam	1 h	60
A142	B2	C2	P2	25%	Written Exam	1 h	60
A143	B3	C3	P3	25%	Written Exam	1 h	60
A144	Controlled Assessment			25%		9 h	64

Command words

The list below explains some of the common words you will see used in exam questions.

Calculate

Work out a number. You can use your calculator to help you. You may need to use an equation. The question will say if your working must be shown. (Hint: don't confuse with 'Estimate' or 'Predict'.)

Compare

Write about the similarities and differences between two things.

Describe

Write a detailed answer that covers what happens, when it happens, and where it happens. Talk about facts and characteristics. (Hint: don't confuse with 'Explain'.)

Discuss

Write about the issues related to a topic. You may need to talk about the opposing sides of a debate, and you may need to show the difference between ideas, opinions, and facts.

Estimate

Suggest an approximate (rough) value, without performing a full calculation or an accurate measurement. Don't just guess – use your knowledge of science to suggest a realistic value. (Hint: don't confuse with 'Calculate' and 'Predict'.)

Explain

Write a detailed answer that covers how and why a thing happens. Talk about mechanisms and reasons. (Hint: don't confuse with 'Describe'.)

Evaluate

You will be given some facts, data, or other kind of information. Write about the data or facts and provide your own conclusion or opinion on them.

Justify

Give some evidence or write down an explanation to tell the examiner why you gave an answer.

Outline

Give only the key facts of the topic. You may need to set out the steps of a procedure or process – make sure you write down the steps in the correct order.

Predict

Look at some data and suggest a realistic value or outcome. You may use a calculation to help. Don't guess – look at trends in the data and use your knowledge of science. (Hint: don't confuse with 'Calculate' or 'Estimate'.)

Show

Write down the details, steps, or calculations needed to prove an answer that you have given.

Suggest

Think about what you've learnt and apply it to a new situation or context. Use what you have learnt to suggest sensible answers to the question.

Write down

Give a short answer, without a supporting argument.

Top Tips

Always read exam questions carefully, even if you recognise the word used. Look at the information in the question and the number of answer lines to see how much detail the examiner is looking for.

You can use bullet points or a diagram if it helps your answer.

If a number needs units you should include them, unless the units are already given on the answer line.

Making sense of graphs

Scientists use graphs and charts to present data clearly and to look for patterns in the data. You will need to plot graphs or draw charts to present data and then describe and explain what the data is showing. Examination questions may also give you a graph and ask you to describe and explain what a graph is telling you.

Reading the axes

Look at these two charts, which both provide data about daily energy use in several countries.

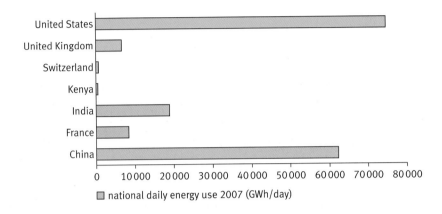

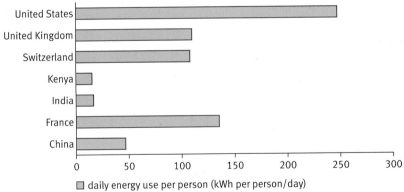

Graphs to show energy use in a range of countries, total and per capita.

Why are the charts so different if they both represent information about energy use?

Look at the labels on the axes.

One shows the **energy use per person per day**, the other shows the **energy use per day by the whole country**.

For example, the first graph shows that China uses a similar amount of energy to the US. But the population of China is much greater – so the energy use per person is much less.

First rule of reading graphs: read the axes and check the units.

Describing the relationship between variables

The pattern of points plotted on a graph shows whether two **factors** are related. Look at this scatter graph.

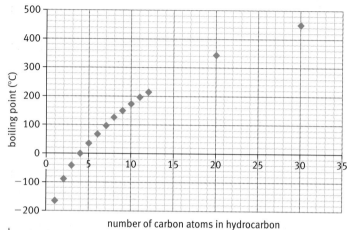

Graph to show the relationship between the number of carbon atoms in a hydrocarbon and the boiling point.

There *is* a pattern in the data; as the number of carbon atoms increases, the boiling point increases.

But it is not a straight line, it is quite a smooth curve, so we can say more than that. When the number of carbon atoms is small the boiling point increases quickly with each extra carbon atom. As the number of carbon atoms gets bigger, the boiling point still increases, but less quickly. Another way of describing this is to say that the slope of the graph – the **gradient** – gets less as the number of carbon atoms increases.

Look at the graph on the right, which shows how the number of bacteria infecting a patient changes over time.

How many different gradients can you see?

There are three phases to the graph, each with a different gradient. So you should describe each phase, including **data** if possible:

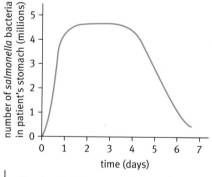

Graph of bacteria population against time.

- The number of bacteria **increases rapidly** for the first day until there are about **4.5 million** bacteria.
- For about **the next three days** the number remains steady at about 4.5 million.
- After the **fourth** day the number of bacteria declines to less than a **million** over the following **two to three days**.

Second rule of reading graphs: describe each phase of the graph, and include ideas about the **gradient** and **data**, including **units**.

Is there a correlation?

Sometimes we are interested in whether one thing changes when another does. If a change in one factor goes together with a change in something else, we say that the two things are **correlated**.

The two graphs on the right show how global temperatures have changed over time and how levels of carbon dioxide in the atmosphere have changed over time.

Is there a correlation between the two sets of data?

Look at the graphs – why is it difficult to decide if there is a correlation?

The two sets of data are over different periods of time, so although both graphs show a rise with time, it is difficult to see if there is a correlation.

It would be easier to identify a correlation if both sets of data were plotted for the same time period and placed one above the other, or on the same axes, like this:

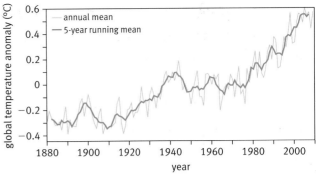

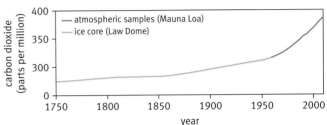

Graphs to show increasing global temperatures and carbon dioxide levels. Source: NASA.

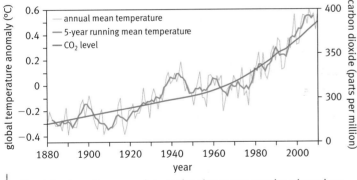

Graph to show the same data as the above two graphs, plotted on one set of axes.

When there are two sets of data on the same axes take care to look at which axis relates to which line.

Third rule for reading graphs: when looking for a correlation between two sets of data, read the axes carefully.

Explaining graphs

When a graph shows that there is a correlation between two sets of data, scientists try to find out if a change in one factor causes a change in the other. They use science ideas to look for an underlying mechanism to explain why two factors are related.

Controlled assessment

In GCSE Science the controlled assessment counts for 25% of your total grade. Marks are given for a case study and a practical data analysis task.

Your school or college may give you the mark schemes for this.

This will help you understand how to get the most credit for your work.

Tip

The best advice is 'plan ahead'. Give your work the time it needs and work steadily and evenly over the time you are given. Your deadlines will come all too quickly, especially if you have coursework to do in other subjects.

Case study (12.5%)

Everyday life has many questions science can help to answer. You may meet these in media reports, for example, on television, radio, in newspapers, and in magazines. A case study is a report that weighs up evidence about a scientific question.

OCR will provide a news sheet with a variety of articles about some of the science topics you have studied in this course.

You will choose an issue from the news sheet as the basis for your case study, and identify a question that you can go on to answer. Your question will probably fit into one of these categories:

- a question where the scientific knowledge is not certain, for example, 'Does using mobile phones cause brain damage?'
- a question about decision making using scientific information, for example, 'Should cars be banned from a shopping street to reduce air pollution?' or 'Should the government stop research into human cloning?'
- a question about a personal issue involving science, for example, 'Should my child have the MMR vaccine?'

You should find out what different people have said about the issue. Then evaluate this information and reach your own conclusion.

You will be awarded marks for:

Selecting information

- Collect information from different places – books, the Internet, newspapers.
- Say where your information has come from.
- Choose only information that is relevant to the question you are studying.
- Decide how reliable each source of information is.

Understanding the question

- Use scientific knowledge and understanding to explain the topic you are studying.
- When you report what other people have said, say what scientific evidence they used (from experiments, surveys, etc.).

Reaching your own conclusion

- Compare different evidence and points of view.
- Consider the benefits and risks of different courses of action.
- Say what you think should be done, and link this to the evidence you have reported.

Presenting your study

- Make sure your report is laid out clearly in a sensible order – use a table of contents to help organise your ideas.
- You may use different presentation styles, for example, a written report, newspaper article, PowerPoint presentation, poster or booklet, or web page.
- Use pictures, tables, charts, graphs, and so on to present information.
- Take care with your spelling, grammar, and punctuation, and use scientific terms where they are appropriate.

Creating a case study

Where do I start?

Read the news sheet you are given and think of a question you want to find the answer to.

Sources of information could include:
- Internet
- school library
- local public library
- your science textbook and notes
- TV
- radio
- newspapers and magazines
- museums and exhibitions.

When will I do my controlled assessment?

Your case study will be written in class time over a series of lessons.

You may also do some research out of class.

Your practical data analysis task will be done in class time over a series of lessons.

Your school or college will decide when you do your controlled assessment. If you do more than one case study or practical data analysis, they will choose the one with the best marks.

Practical data analysis (12.5%)

Scientists collect data from experiments and studies. They use this data to explain how something happens. You need to be able to assess the methods and data from scientific experiments. This will help you decide how reliable a scientific claim is.

A practical data analysis task is based on a practical experiment that you carry out. The experiment will be designed to test a hypothesis suggested by your teacher. You may do the experiment alone or work in groups and pool all your data. Then you interpret and evaluate the data.

You will be awarded marks for:

Choosing how to collect the data
- Carry out the experiment in ways that will give you high-quality data.
- Explain why you chose this method.
- Explain how you worked safely.

Interpreting data
- Present your data in tables, charts, or graphs.
- Say what conclusions you can reach from your data.
- Explain your conclusions using your scientific knowledge and understanding.

Evaluating the method and quality of data
- Look back at your experiment and say how you could improve the method.
- Explain how confident you are in your evidence. Have you got enough results? Do they show a clear pattern? Have you repeated measurements to check them? Would you get the same results if you repeated the experiment?
- Comment on the repeatability of your data, account for any outliers in the data, or explain why there are no outliers.
- Suggest some improvements or extra data you could collect to be more confident in your conclusions.

Reviewing the hypothesis
- Use your scientific knowledge to decide whether the hypothesis is supported by your data.
- Suggest what extra data could be collected to increase confidence in the hypothesis.

Presenting your report
- Make sure your report is laid out clearly in a sensible order.
- Use diagrams, tables, charts, and graphs to present information.
- Take care with your spelling, grammar, and punctuation, and use scientific terms where they are appropriate.

B1 You and your genes

Why study genes?

What makes me the way that I am? Your ancestors probably asked the same question. How are features passed on from parents to children? You may look like your relatives, but you are unique. Only in the last few generations has science been able to answer questions like these.

What you already know

- In sexual reproduction fertilisation happens when a male and female sex cell join together. Information from two parents is mixed to make a new plan for the offspring. The offspring will be similar but not identical to their parents.

- There are variations between members of the same species that are due to environmental as well as inherited causes.

- Clones are individuals with indentical genetic information.

- The science of cloning raises ethical issues.

Find out about

- how genes and your environment make you unique

- how and why people find out about their genes

- how we can use our knowledge of genes

- whether we should allow this.

The Science

Your environment has a huge effect on you, for example, on your appearance, your body, and your health. But these features are also affected by your genes. In this Module you'll find out how. You'll discover the story of inheritance.

Ideas about Science

In the future, science could help you to change your baby's genes before it is born. Cloned embryos could provide cells to cure diseases. But, as new technologies are developed, we must decide how they should be used. These can be questions of ethics – decisions about what is right and wrong.

Find out about

- ✓ **what makes us all different**
- ✓ **what genes are and what genes do**

Plants and animals look a lot like their parents. They have **inherited** information from them. This information is in **genes** and controls how the organisms develop and function.

A lot of information goes into making a human being. So inheritance does a big job pretty well. All people have most features in common. Children look a lot like their parents. If you look at the people around you, the differences between us are very, very small. But we're interested in them because they make us unique.

Most features are affected by both the information you inherit and your environment.

These sisters have some features in common.

Environment makes a difference

Almost all of your features are affected by the information you inherited from your parents. For example, your blood group depends on this information. Some features are the result of only your environment, such as scars and tattoos.

But most of your features are affected by both your genes and your **environment**. For example, For example, your weight depends on inherited information, but if you eat too much, you become heavier

---

(handwritten notes)

① Skin colour, eye colour, height, hair colour, curly/straught hair.

② environment & genes

③ inherited information is in genes and controls how the organism develops and functions. All information goes into making a human being.

- ✓ genes
- ✓ environment

Questions

1 Choose two of the students in the photograph on the left. Write down five ways they look different.

2 What two things can affect how you develop?

3 Explain what is meant by inherited information.

Inheritance – the story of life

One important part of this story is where all the information is kept. Living organisms are made up of cells. If you look at cells under a microscope you can see the **nuclei**. Inside each nucleus are long threads called **chromosomes**. Each chromosome contains thousands of genes. I is genes that control how you develop.

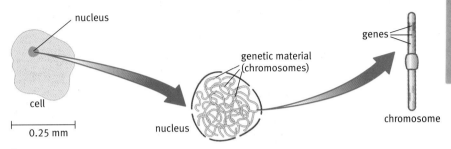

nucleus

cell

0.25 mm

genetic material
(chromosomes)

nucleus

genes

chromosome

All the information needed to create a whole human being fits into the nucleus of a cell. The nucleus is just 0.006 mm across!

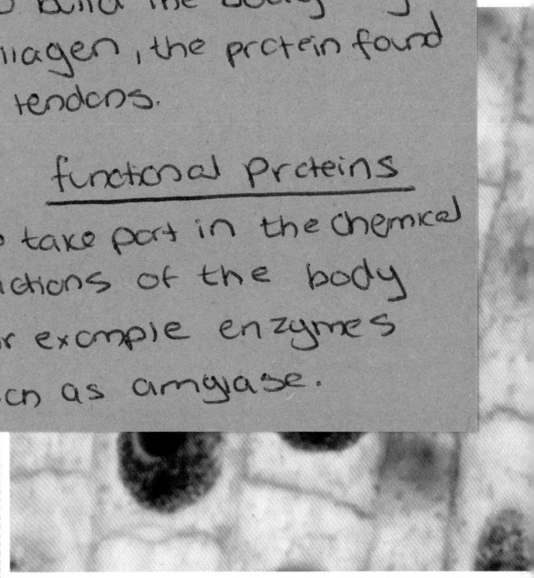

What are chromosomes made of?

Chromosomes are made of very long molecules of **DNA**. DNA is short for deoxyribonucleic acid. A gene is a section of a DNA molecule.

How do genes control your development?

A fertilised egg cell has the instructions for making every **protein** in a human being. That's what genes are – instructions for making proteins. Each gene is the instruction for making a different protein.

What's so important about proteins?

Proteins are important chemicals for cells. There are many different proteins in the body, and each one does a different job. They may be:

- **structural** proteins – to build the body, eg collagen, the protein found in tendons
- **functional** proteins – to take part in the chemical reactions of the body, for example, enzymes such as amylase.

Genes control which proteins a cell makes. This is how they control what the cell does and how an organism develops.

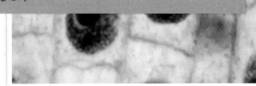

These plant cells have been stained to show up their nuclei. One cell is dividing and the separating chromosomes can be seen.

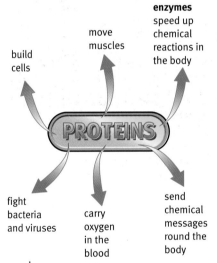

enzymes speed up chemical reactions in the body

move muscles

build cells

PROTEINS

fight bacteria and viruses

carry oxygen in the blood

send chemical messages round the body

There are about 50 000 types of

Questions

4   Write these cell parts in order starting with the smallest: chromosome, gene, cell, nucleus

5   Explain how genes control what a cell does.

6   a   List two kinds of job that proteins do in the human body.
    b   Name two proteins in the human body and say what they do.

Handwritten notes:

Structural PROTEINS

TO build the body e.g collagen, the protein found in tendons.

functional Proteins

to take part in the chemical reactions of the body for example enzymes such as amylase.

④ gene, chromosome, nucleus, cell (smallest → biggest)

⑤ Genes control what does which proteins a cell makes. This is how this is how they control what the cell does and how an organism develops.

⑥a build the body, to take part in the chemical reactions of the body.

Find out about

- ✔ why identical twins look like each other
- ✔ why identical twins do not stay identical
- ✔ what a clone is

When a baby is born, who can say how it will grow and develop? Your genes decide a lot about you. A few characteristics, like having dangly earlobes or dimples, are decided by just one pair of genes. Most of your characteristics, such as your height, your weight, and your eye colour, are decided by several different genes working together. But your genes don't tell the whole story.

Twins and the environment

① Dimples can happen if the person inherits a D allele from each parent or from one. It is more dominent. Being 2m tall can also be effected by your environment as well as your genes.

② They find it useful to see whether or not the environment has a 8 massive affect on our appecronce. It allows scientists to find out how much of our characteristics are because of genes and how much by our environment

Identical twins have the same genes but they don't look exactly the same.

Identical twins are formed when a fertilised egg starts to divide and splits to form two babies instead of one. Each baby has the same genetic information. This means any differences between them are because of the environment.

Most identical twins are brought up together in the same family so their environment is very similar. But sometimes twins are separated after birth and adopted by different parents. This allows scientists to find out how much their characteristics are because of their genes, and how much they are affected by the different environments they live in.

It is often surprising how alike the separated twins are when they meet. The influence of the genes is very strong. The different environments do mean that some things, like the weights of the twins, are more different than for twins who are brought up together.

Identical twins that are separated at birth are not very common. It would be wrong for scientists to separate babies just to find out how much effect the environment has on how they grow and develop. For this reason scientists often study plants instead.

Questions

1 What are the differences between the ways that having dimples, having green eyes, and being 2 m tall are determined?

2 Why do scientists find studying identical twins so useful?

Cloning

We call any genetically identical organisms **clones**. So identical twins are human clones! Plant clones are quite common. Strawberry plants and spider plants are just two sorts of plant that make identical plant clones at the end of runners. Bulbs, like daffodils, also produce clones.

It is easy for people to clone plants artificially. This can be done by taking cuttings. A piece of the adult plant is cut off. It soon forms new roots and stems to become a small plant genetically identical to the original parent. The new plant is a clone – it has the same genes as the parent plant.

You can also make clones from tiny pieces of plants grown in special jelly, called agar. In this way you can make hundreds of clones from a single plant.

Once you have some clones, you can look at how the environment affects them. If the parent plant grew very tall, that will be partly down to its genes. But what happens if it doesn't get enough nutrients, or it is short of water? Will it still grow tall? When we look at the effect of different factors on the characteristics of cloned plants, it helps us to understand how genes and the environment interact.

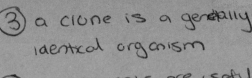

③ a clone is a genetically identical organism

④ cloned plants are useful to scientists because you can look at how the enviroment effets them or for e.g if one doesnt get enough nutrients would it change the plant? will it grow different?

Each of these baby spider plants is a clone of its parent plant and of all the other baby plants.

You may make cauliflower clones like these.

Questions

3 What is a clone?

4 Why are cloned plants so useful to scientists?

5 Describe how you could use cloned plants to show how the environment affects their appearance.

⑤ you could use cloned plants by giving one plant something and deprive the other plant of it so for e.g the amount of water given. You can then compare the plant once identical plants and see what features are different or if there are any changes.

It can be funny to see that people in a family look like each other. Perhaps you don't like a feature you've inherited – your dad's big ears or your mum's freckles. For some people, family likenesses are very serious.

Craig's story

My grandfather's only 56. He's always been well but now he's a bit off colour. He's been forgetting things – driving my Nan mad. No-one's said anything to me, but they're all worried about him.

Robert's story

I'm so frustrated with myself. I can't sit still in a chair. I'm getting more and more forgetful. Now I've started falling over for no reason at all. The doctor has said it might be **Huntington's disease**. It's an inherited disorder. She said I can have a blood test to find out, but I'm very worried.

Huntington's disease

Huntington's disease is an inherited disorder. You can't catch it. The disease is passed on from parents to their children. The symptoms of Huntington's disease don't appear until middle age. First the person has problems controlling their muscles. This shows up as twitching. Gradually a sufferer becomes forgetful. They find it harder to understand things and to concentrate. People with Huntington's often have mood changes. After a few years they can't control their movements. Sadly, the condition is fatal.

① memory loss
loose control of muscles
harder to understand &
concentrate.

② It is called an inherited
disorder because you
cannot catch it. Too
It can only be passed
on from parents to their
children.

Robert, 56
I've been forgetting things and stumbling.

Eileen, 58
Robert's mum was just the same. David looks just like his father.

Sarah, 32
I'm definitely having the test if Dad's got it. I need to know so I can plan my life.

David, 35
I'm not having a test. It won't change what happens to me.

Clare, 33
David's got the right idea, just getting on with his life. Mind you, I'm really worried about him now – and Craig and Hannah.

Craig, 16
It's not fair. I want to find out but they won't let me. They think I'm too young to understand.

Hannah, 14
No-one seems to want to tell me anything about it at all.

Craig's family tree.

Questions

1 List the symptoms of Huntington's disease.

2 Explain why Huntington's disease is called an inherited disorder.

How do you inherit your gen[...]

Sometimes people in the same family look a lot alike. In other families brothers and sisters look very different. They may also look different from their parents. The key to this mystery lies in our genes.

Parents pass on genes in their **sex cells**. In animals these are sperm and egg cells (ova). Sex cells have copies of half the parent's chromosomes. When a sperm cell fertilises an egg cell, the fertilised egg cell gets a full set of chromosomes. It is called an **embryo**.

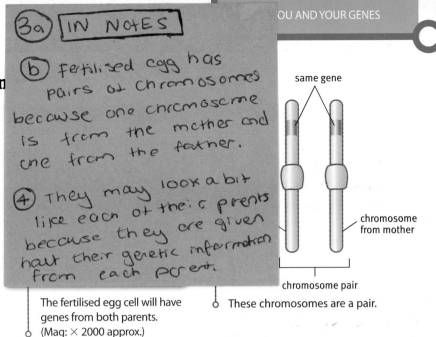

(3a) IN NOTES

(b) Fertilised egg has pairs of chromosomes because one chromosome is from the mother and one from the father.

(4) They may look a bit like each of their parents because they are given haut their genetic information from each parent.

same gene

chromosome from mother

chromosome pair

These chromosomes are a pair.

The fertilised egg cell will have genes from both parents. (Mag: × 2000 approx.)

How many chromosomes does each cell have?

Chromosomes come in pairs. Every human body cell has 23 pairs of chromosomes. The chromosomes in most pairs are the same size and shape. They carry the same genes in the same place. This means that your genes also come in pairs.

Sex cells have single chromosomes

Sex cells are made with copies of half the parent's chromosomes, one from each pair. This makes sure that the fertilised egg cell has the right number of chromosomes – 23 pairs.

One chromosome from each pair came from the egg cell. The other came from the sperm cell. Each chromosome carries thousands of genes. Each chromosome in a pair carries the same genes along its length.

So the fertilised egg cell has a mixture of the parents' genes. Half of the new baby's genes are from the mother. Half are from the father. This is why children resemble both their parents.

(5) because the genetic information the each get will be in a random order so they will not inherit the same genetic information.

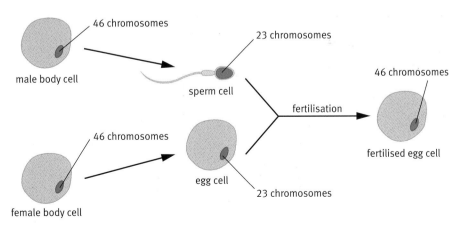

46 chromosomes

male body cell

sperm cell

23 chromosomes

fertilisation

46 chromosomes

fertilised egg cell

46 chromosomes

female body cell

egg cell

23 chromosomes

The cells in this diagram are not drawn to scale. A human egg cell is 0.1 mm across. This is 20 times larger than a human sperm cell.

Questions

3 a Draw a diagram to show a sperm cell, an egg cell, and the fertilised egg cell they make.

b Explain why the fertilised egg cell has pairs of chromosomes.

4 Explain why children may look a bit like each of their parents.

5 Two sisters with the same parents won't look exactly alike. Explain why this is.

Find out about

- ✔ **what decides if you are male or female**
- ✔ **how a Y chromosome makes a baby male**

Ever wondered what it would be like to be the opposite sex? Well, if you are male there was a time when you were female – just for a short while. Male and female embryos are very alike until they are about six weeks old.

What decides an embryo's sex?

A fertilised human egg cell has 23 pairs of chromosomes. Pair 23 are the sex chromosomes. Males have an X chromosome and a Y chromosome – **XY**. Females have two X chromosomes – **XX**.

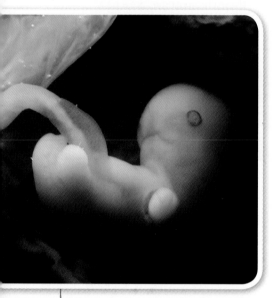

This embryo is six weeks old.

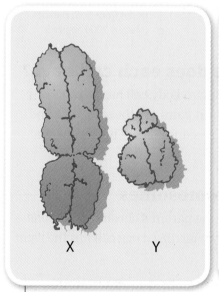

Women have two X chromosomes. Men have an X and a Y.

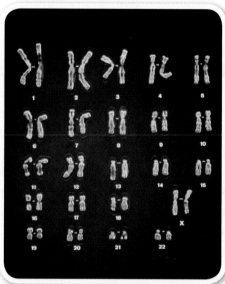

These chromosomes are from the nucleus of a woman's body cell. They have been lined up to show the pairs.

Key words

- ✔ **XY chromosomes**
- ✔ **XX chromosomes**

What's the chance of being male or female?

A parent's chromosomes are in pairs. When sex cells are made they only get one chromosome from each pair. So half a man's sperm cells get an X chromosome and half get a Y chromosome. All a woman's egg cells get an X chromosome.

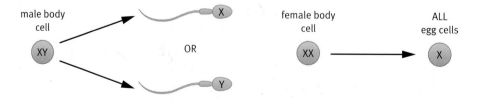

When a sperm cell fertilises an egg cell the chances are 50% that it will be an X or a Y sperm. This means that there is a 50% chance that the baby will be a boy or a girl.

How does the Y chromosome make a baby male?

A male embryo's testes develop when it is about six weeks old. This is caused by a gene on the Y chromosome – the SRY gene. SRY stands for 'sex-determining region of the Y chromosome'.

Testes produce male sex hormones called androgens. Androgens make the embryo develop into a male. If there is no male sex hormone present, the sex organs develop into the ovaries, clitoris, and vagina of a female.

SRY gene

The Y chromosome

What are hormones?

Hormones are chemicals that control many processes in the cells. Tiny amounts of hormones are made by different parts of the body. You can read more about hormones in Module B2: *Keeping Healthy*.

Jan's story

At eighteen Jan was studying at college. She was very happy, and was going out with a college football player. She thought her periods hadn't started because she did a lot of sport.

Then in a science class Jan looked at the chromosomes in her cheek cells. She discovered that she had male sex chromosomes – XY.

Sometimes a person has X and Y chromosomes but looks female. This is because their body makes androgens but the cells take no notice of it. About 1 in 20000 people have this condition. They have small internal testes and a short vagina. They can't have children.

Jan had no idea she had this condition. She found it very difficult to come to terms with. But she has now told her boyfriend and they have stayed together.

1a) XY
b) X
c) XX
d) X or Y

2) It needs a Y chromosome because It contains a SRY gene. This produces a hormone called androgen & this makes the embryo develop into a male.

Questions

1 Which sex chromosome(s) would be in the nucleus of:
   a a man's body cell?
   b an egg cell?
   c a woman's body cell?
   d a sperm cell?

2 Explain why an embryo needs the Y chromosome to become a boy.

3 Imagine you are Jan or her boyfriend. How would you have felt about her condition?

Jan on holiday, aged eighteen.

Find out about

- ✔ **how pairs of genes control some features**
- ✔ **cystic fibrosis (an inherited illness)**
- ✔ **testing a baby's genes before they are born**

Key words

- ✔ alleles
- ✔ dominant
- ✔ recessive
- ✔ phenotype
- ✔ genotype

This diagram shows one pair of chromosomes. The gene controlling dimples is coloured in.

Will this baby be tall and have red hair? Will she have a talent for music, sport – even science?! Most of these features will be affected by both her environment and her genes. A few features are controlled by just one gene. We can understand these more easily.

This baby has inherited a unique mix of genetic information.

Genes come in different versions

Both chromosomes in a pair carry genes that control the same features. Each pair of genes is in the same place on the chromosomes.

But genes in a pair may not be exactly the same. They may be slightly different versions. You can think about it like football strips – often a team's home and away strips are both based on the same pattern, but they're slightly different. Different versions of the same genes are called **alleles**.

Dominant alleles – they're in charge

The gene that controls dimples has two alleles. The D allele gives you dimples. The d allele won't cause dimples. The alleles you inherit is called your **genotype**.

Your **phenotype** is what you look like, eg dimples or no dimples, your characteristics.

The D allele is **dominant**. You only need one copy of a dominant allele to have its feature. The d allele is **recessive**. You must have two copies of a recessive allele to have its feature – in this case no dimples.

Which alleles can a person inherit?

Sex cells get one chromosome from each pair the parent has. So they only have one allele from each pair. If a parent has two D or two d alleles, they can only pass on a D or a d allele to their children.

dimples

This person inherited a D allele from both parents. They have dimples.

no dimples

This person inherited a d allele from both parents. They don't have dimples.

dimples

This person inherited one D and one d allele. They have dimples.

But a parent could have one D and one d allele. Then half of their sex cells will get the D allele and half will get the d allele.

The human lottery

We cannot predict which egg and sperm cells will meet at fertilisation. This genetic diagram (Punnett square) shows all the possibilities for one couple.

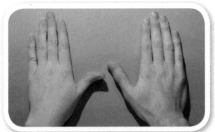

The allele that gives you straight thumbs is dominant (T). The allele for curved thumbs is recessive (t).

The allele that gives you hair on the middle of your fingers is dominant (R). The allele for no hair is recessive (r).

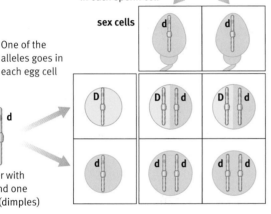

A father with two d alleles (no dimples)

One of the alleles goes in each sperm cell

sex cells

One of the alleles goes in each egg cell

mother

A mother with one D and one d allele (dimples)

children There is a 50% chance of a child having dimples

Why don't brothers and sisters look the same?

Human beings have about 23 000 genes. Each gene has different alleles. If both of the alleles you inherit are the same you are **homozygous** for that characteristic. If you inherit different alleles you are **heterozygous**.

Brothers and sisters are different because they each get a different mixture of alleles from their parents. Except for identical twins, each one of us has a unique set of genes.

What about the family?

Huntington's disease is a single-gene disorder caused by a dominant allele. You only need to inherit the allele from one parent to have the condition. In Craig's family, whom we met on page 20, Craig's grandfather, Robert, has Huntington's disease. So their dad, David, may have inherited this faulty allele. At the moment he has decided not to have the test to find out.

Questions

1 Write down what is meant by the word allele.

2 Explain how you inherit two alleles for each gene.

3 Explain the difference between a dominant and a recessive allele.

4 What are the possible pairs of alleles a person could have for:
   a dimples?
   b straight thumbs?
   c no hair on the second part of their ring finger?

5 Use a diagram to explain why a couple who have dimples could have a child with no dimples.

6 Use a diagram to work out the chance that David has inherited the Huntington's disease allele.

Dear Clare,

Please help us. My husband Huw and I have just been told that our first child has cystic fibrosis. No one in our family has ever had this disease before. Did I do something wrong during my pregnancy? I'm so worried.

Yours sincerely
Emma

Dear Emma,

What a difficult time for you all. First of all, nothing you did during your pregnancy could have affected this, so don't feel guilty. Cystic fibrosis is an inherited disorder ...

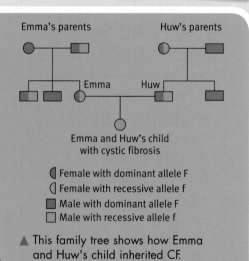

Emma's parents   Huw's parents

Emma   Huw

Emma and Huw's child
with cystic fibrosis

◖ Female with dominant allele F
◖ Female with recessive allele f
◼ Male with dominant allele F
◻ Male with recessive allele f

▲ This family tree shows how Emma and Huw's child inherited CF.

Key words

- ✓ cystic fibrosis
- ✓ termination
- ✓ carrier

Dear Doctor

We've had a huge postbag in response to last month's letter from Emma. So this month we're looking in depth at cystic fibrosis, a disease that one in twenty-five of us carries in the UK ...

What is cystic fibrosis?

You can't catch cystic fibrosis. It is a genetic disorder. This means it is passed on from parents to their children. The disease causes big problems for breathing and digestion. Cells that make mucus in the body don't work properly. The mucus is much thicker than it should be, so it blocks up the lungs. It also blocks tubes that take enzymes from the pancreas to the gut. People with cystic fibrosis get breathless. They also get lots of chest infections. The shortage of enzymes in their gut means that their food isn't digested properly. So the person can be short of nutrients.

How do you get cystic fibrosis?

Most people with cystic fibrosis (CF) can't have children. The thick mucus affects their reproductive systems. So babies with CF are usually born to healthy parents. At first glance this seems very strange – how can a parent pass a disease on to their children when they don't have it themselves?

The answer lies with one of the thousands of genes responsible for producing a human being. One of these instructs cells to make mucus. But sometimes there are errors in the DNA, so that it does not do its job.

A person who has one dominant, normal-functioning allele (F) and one recessive, faulty allele (f) will not have CF. They can still make normal mucus. But they are a carrier of the faulty allele. When parents who are carriers make sex cells, half will contain the normal allele – and half will contain the faulty allele. When two sex cells carrying the faulty allele meet at fertilisation, the baby will have CF. One in every 25 people in the UK carries the CF allele.

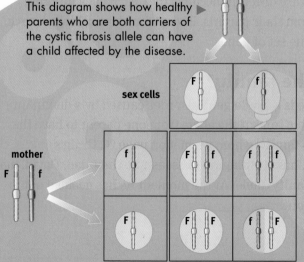

This diagram shows how healthy ▶ parents who are both carriers of the cystic fibrosis allele can have a child affected by the disease.

F   f father

sex cells

mother
F   f

children There is a 25% chance that a child from the carrier parents will have cystic fibrosis.

Can cystic fibrosis be cured?

Not yet. But treatments are getting better, and life expectancy is increasing all the time. Physiotherapy helps to clear mucus from the lungs. Sufferers take tablets with the missing gut enzymes in. Antibiotics are used to treat chest infections. And an enzyme spray can be used to thin the mucus in the lungs, so it is easier to get rid of. New techniques may offer hope for a cure in the future.

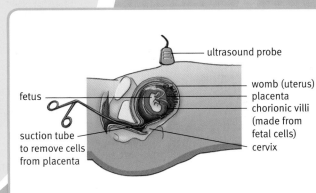

▲ Tom has cystic fibrosis. He has physiotherapy every day to clear thick mucus from his lungs.

What are the options?

If a couple know they are at risk of having a child with cystic fibrosis they can have tests to see if their child has the disease. During pregnancy cells from the developing fetus can be collected, and the genes examined. If the fetus has two faulty alleles for cystic fibrosis, the child will have the disease. The parents may choose to end the pregnancy. This is done with a medical operation called a termination (an abortion).

How do doctors get cells from the fetus?

The fetal cells can be collected two ways:
- an amniocentesis test
- a chorionic villus test.

The diagrams show how each of these tests is carried out.

▼ Amniocentesis test.

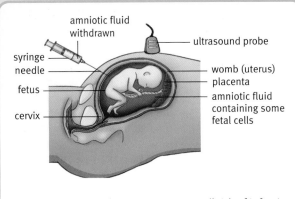

amniotic fluid withdrawn

ultrasound probe

syringe needle

fetus

cervix

womb (uterus)
placenta
amniotic fluid containing some fetal cells

- 1% miscarriage risk
- results at 15–18 weeks
- very small risk of infection
- results not 100% reliable

▼ Chorionic villus test.

ultrasound probe

fetus

suction tube to remove cells from placenta

womb (uterus)
placenta
chorionic villi (made from fetal cells)
cervix

- 2% miscarriage risk
- results at 10–12 weeks
- almost no risk of infection
- results not 100% reliable

Questions

7   The magazine doctor is sure that nothing Emma did during her pregnancy caused her baby to have cystic fibrosis. How can she be so sure?

8   People with cystic fibrosis make thick, sticky mucus. Describe the health problems that this may cause.

9   Explain what it means when someone is a 'carrier' of cystic fibrosis.

10 Two carriers of cystic fibrosis plan to have children. Draw a diagram to work out the chance that they will have:
    a   a child with cystic fibrosis
    b   a child who is a carrier of cystic fibrosis
    c   a child who has no cystic fibrosis alleles.

'We had an amniocentesis test for each of my pregnancies,' says Elaine. 'Sadly we felt we had to terminate the first one, because the fetus had CF. We are lucky enough now to have two healthy children – and we know we don't have to watch them suffer.'

Elaine's nephew has cystic fibrosis. When they found out, Elaine and her husband Peter became worried about any children they might have. They both had a **genetic test**. The tests showed that they were both carriers for cystic fibrosis. Elaine and Peter decided to have an amniocentesis test when Elaine was pregnant.

Elaine and Peter made a very hard decision when they decided to terminate their first pregnancy. When a person has to make a decision about what is the right or wrong way to behave, they are thinking about **ethics**. Deciding whether to have a termination is an example of an ethical question.

Ethics – right and wrong

For some ethical questions, the right answer is very clear. For example, should you feed and care for your pet? Of course. But in some situations, like Elaine and Peter's, people may not agree on one right answer. People think about ethical questions in different ways.

For example, Elaine and Peter felt that they had weighed up the consequences of either choice. They thought about how each choice – continuing with the pregnancy or having a termination – would affect all the people involved. They had to make a judgement about the difficulties their unborn child would face with cystic fibrosis.

In order to consider all the consequences they also had to think about the effects that an ill child would have on their lives, and on the lives of any other children they might have. Some people feel that they could not cope with the extra responsibility of caring for a child with a serious genetic disorder.

Different choices

Not everyone weighing up the consequences of each choice would have come to the same decision as this couple did.

Jo has a serious genetic disorder. Her parents believe that termination is wrong. They decided not to have more children, rather than use information from an amniocentesis test.

Some people feel that any illness would have a devastating effect on a person's quality of life. But some people lead very happy, full lives in spite of very serious disabilities.

Elaine and Peter made their ethical decision only by thinking about the consequences that each choice would have. This is just one way of dealing with ethical questions.

When you believe that an action is wrong

For some people having a termination would be completely wrong in itself. They believe that an unborn child has the right to life, and should be protected from harm in the same way as people are protected after they are born. Other people believe that terminating a pregnancy is unnatural, and that we should not interfere. People may hold either of these viewpoints because of their own personal beliefs, or because of their religious beliefs.

A couple in Elaine and Peter's position who felt that termination was wrong might decide not to have children at all. This would mean that they could not pass on the faulty allele. Or they could decide to have children, and to care for any child that did inherit the disease.

What are the ethical arguments for a decision?

The right decision is the one which leads to the best outcome for the most people.

Some actions are wrong and should never be done.

It's wrong to have a termination. We'll look after our baby whatever.

Is it fair for us to have a baby knowing they'll be ill?

Questions

1 Explain what is meant by 'an ethical question'.

2 Describe three different points of view that a couple in Elaine and Peter's position might take.

3 What is your viewpoint on genetic testing of a fetus for a serious illness? Explain why you think this.

Find out about

✓ **what a genetic test is**
✓ **what genetic screening is**

How reliable are genetic tests?

Genetic testing is used to look for alleles that cause genetic disorders. People like Elaine and Peter use this information to make decisions about whether to have children. Genetic tests can also be used to make a decision about whether a pregnancy should be continued or not.

So, it is important to realise that the tests are not completely reliable. In a few cases only it will not detect CF. This is called a **false negative**. The test only looks for common DNA errors in the faulty CF gene. **False positive** tests are not common, but they may happen due to technical failure of the test.

Why do people have genetic tests?

Usually people only have a genetic test because they know that a genetic disorder runs in their family; they want to know if they are carriers. Most parents who have a child with cystic fibrosis did not know that they were carriers. So, they would not have had a genetic test during pregnancy.

Every newborn baby in the UK is now screened for cystic fibrosis. They have a blood test – this is a biochemical test, not the more expensive genetic test. If the biochemical test is positive for CF the baby will be genetically tested to confirm the diagnosis. Treatment can start before the lungs are too badly damaged. Gentetic testing the whole population or large groups for a genetic disease is called **genetic screening**.

This couple are both carriers of cystic fibrosis. They had an amniocentesis test during their pregnancy. The results showed that the baby did not have CF. When their daughter was born she was completely healthy.

Who decides about genetic screening?

The decision about whether to use genetic screening is taken by governments and local NHS trusts. People in the NHS have to think about different things when they decide if genetic screening should be used:

- What are the costs of testing everyone?
- What are the benefits of testing everyone?
- Is it better to spend the money on other things, for example, hip replacement operations, treating people with cancer, and treating people who already have cystic fibrosis?

Can we? Should we?

Four of Rabbi Joseph Ekstein's children died from a severe genetic condition called Tay-Sachs disease. In the general population about one baby in every 300000 has Tay-Sachs, but in the 1980s one baby in every 3600 born to European Jewish families was affected and died.

Key words

✓ **false negative**
✓ **false positive**
✓ **genetic screening**

In 1983 Rabbi Ekstein set up a genetic screening programme. Couples planning to marry were genetically tested. If they both carried the recessive allele they were advised either to not marry or to have prenatal screening and terminate affected pregnancies. As a result of this genetic screening, Tay-Sachs has almost disappeared from Jewish communities worldwide.

Testing your genes

If you could have more information about your genes, would you want it? You can buy a DNA testing kit that tells you if you are carrying faulty alleles for over 100 rare genetic mutations including cystic fibrosis and Tay-Sachs. Some scientists hope tests like these will help to prevent many genetic diseases. Other scientists think that the risk of being affected by these rare genetic diseases is so low that screening is not worthwhile. It costs money and may cause people to worry unnecessarily.

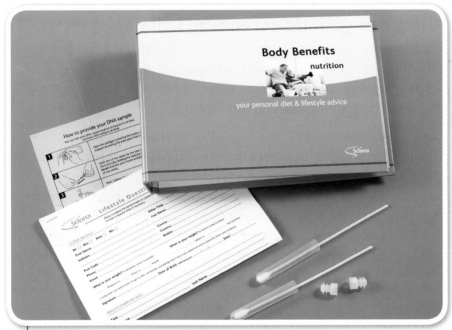

DNA testing can already be done at homes with simple kits like this. How much do you want to know about your DNA?

In the future

Scientists can already work out the complete DNA sequence (the genome) of anyone who has enough money to pay thousands of pounds. In five years' time it may be so cheap that everyone will be able to have it done. The genome of every newborn baby may be worked out. How might we use this information?

Questions

1 What are 'false negative' and 'false positive' results?

2 Why is it important for people to know about false results?

3 Explain what is meant by the term 'genetic screening'.

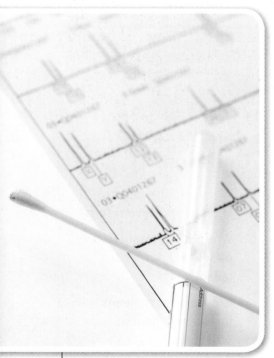

Carolyn had a dangerous reaction to drugs. Genetic testing may help to avoid this.

In the future, decisions about drug treatments may be based on genetic test results.

Finding the right medicine

In 2009 Carolyn Major started to take the medicine she hoped would help cure her cancer. Instead, four days later, she was in an intensive care ward with her heart struggling to keep going. She is one of a small group of people whose bodies react very badly to the anti-cancer drug. Carolyn was lucky. She recovered with no permanent damage done to her heart – and so far the cancer hasn't returned either.

In future the genetic testing of adults may make problems like this a thing of the past. Doctors think that we can use genetic testing to match medicines to patients. Some people produce enzymes that break down drugs very quickly, so they need higher doses of a medicine than most of us. Other people don't produce enzymes that break down certain drugs in their body, so they can be poisoned by medicine that is meant to help them. In future, genetic testing may show if people have the genes for these enzymes so we can all be given the drugs that work best for our bodies' genotype.

What can genetic screening tell us?

Full genetic screening will show if you have inherited genes that increase your risk of a particular problem such as heart disease or different types of cancer. But remember that most of these diseases are affected by both your genes and your lifestyle.

This information could be very helpful. If you found out that you are genetically at a higher than average risk of developing heart disease, for example, you might decide to change your lifestyle. If you don't smoke, eat a healthy diet, and always take lots of exercise, you would lower your environmental risk of heart disease. This would help to balance your increased genetic risk. And if you know that your genotype means you have an increased risk of getting a particular type of cancer, you can be screened regularly to catch the disease as early as possible if it develops.

Questions

4 How can the genetic testing of adults prevent the birth of babies affected with genetic diseases?

5 What do you think are the advantages and disadvantages of testing adults?

6 How might genetic testing make medicines more effective and what problems might it cause?

Is it right to use genetic screening?

It is easy to see why people may want genetic screening:

- when two people decided to have children, they would know if their children were at risk of inheriting the disorder
- when a disease runs in someone's family, they would know if they were going to develop it.

At first glance, genetic screening may seem like the best course of action for everyone. But the best decision for the majority is not always the right decision. There are ethical questions to consider about genetic screening, including:

- Who should know the test results?
- What effect could the test result have on people's future decisions?
- Should people be made to have screening, or should they be able to opt out? Is it right to interfere?

About 1 in 25 people in the UK carry the allele for cystic fibrosis. Some people think that having this information is useful, but there are also good reasons why not everyone agrees. A decision may benefit many people. But it may not be the right decision if it causes a great amount of harm to a few people.

People have different ideas about whether genetic screening for cystic fibrosis would be a good thing.

Questions

7 Give two arguments for and two against genetic screening for cystic fibrosis.

8 Which argument do you agree with? Explain why.

Who should know about your genes?

If you find out that your genotype suggests you have a higher-than-average risk of developing heart disease, who should know about it? Your partner? Your family? Your boss? Your insurance company?

Many people think that only you and your doctor should know information about your genes. They are worried that it could affect a person's job prospects and chances of getting life insurance.

How does life insurance work?

People with life insurance pay a regular sum of money to an insurance company. This is called a premium. In return, when they die, the insurance company pays out an agreed sum of money. People buy life insurance so that there will be money to support their families when they die.

People use information like this to decide the premium each person should pay for insurance. The higher the risk, the bigger the premium.

Condition	Percentage of deaths from each disease caused by smoking in 2007	
	Men	Women
Cancers		
Lung	88	75
Throat and mouth	76	56
Oesophagus	70	63
Heart and circulation diseases	11	15

Should insurance companies know about your genes?

Insurance companies assess what a person's risk is of dying earlier than average. If they believe that the risk is high, they may choose to charge higher premiums than average. Some people think that insurance companies might use the results of genetic tests in the wrong way. Individual people might do the same. But having a gene that means you have a higher risk does not mean that you will definitely die early. Here are some of the arguments:

- Insurers may not insure people if a test shows that they are more likely to get a particular disease. Or they may charge a very high premium.
- Insurers may say that everyone must have genetic tests for many diseases before they can be insured.
- People may not tell insurance companies if they know they have a genetic disorder.
- People may refuse to have a genetic test because they fear that they will not be able to get insurance. They may miss out on medical treatment that could keep them healthy.

In the USA, the 2008 Genetic Information Nondiscrimination Act means insurance companies can't stop someone getting health insurance, or even charge them higher premiums, because of the results of a genetic test. It also means that employers can't use genetic information when they are deciding who should get a job or who should be fired.

In the UK insurance companies agreed not to collect and use genetic information about people until the end of 2014. There are also some concerns about whether the results of genetic testing might affect employment. Politicians are considering bringing in laws to control how information about people's genes can be used. This debate has been going on for 10 years now and no decisions have been made yet!

What about the police?

Scientists already use information about people's DNA to help them solve crimes. They produce DNA profiles from cells left at a crime scene. There is usually only a 1 in 50 million chance of two people having the same DNA profile – unless they are identical twins.

Some people think that there should be a national DNA database recording everyone living in the country. If all babies were to have their genome sequenced at birth, this will be easily available in the future. At the moment only DNA from people who have been arrested or convicted of committing a crime can be stored. Human rights campaigners are against a database for everyone's DNA. They think it is an invasion of privacy and puts innocent people on the same level as criminals.

DNA evidence has been used to prove many people guilty – and others innocent.

Questions

9 If genetic testing shows you have a higher-than-average risk of heart disease, do you think this should affect your application:
   a to be an airline pilot?
   b to take out life insurance?

10 Write down arguments for and against the government setting up a DNA database.

11 Do you agree or disagree with the genetic testing of newborn babies and the setting up of a DNA database? Explain your reasons.

Find out about

- ✔ how new techniques can allow people to select embryos
- ✔ how people think this technology should be used

Sally takes a 'fertility drug' so that she releases several eggs. Fertility drugs contain hormones.

↓

In a small operation, the doctor collects the eggs.

↓

Bob's sperm fertilise the eggs in a Petri dish. This is in vitro fertilisation.

↓

When the embryos reach the eight-cell stage, one cell is removed from each.

↓

The cells are tested for the Huntington's allele. This is called pre-implantation genetic diagnosis (PGD).

↓

Only embryos without the Huntington's allele are implanted in Sally's uterus.

Many people do not agree with termination. If they are at risk of having a child with a genetic disease, they may have decided not to have children. Now doctors can offer them another treatment. It uses in vitro fertilisation (IVF). In this treatment the mother's egg cells are fertilised outside her body. This treatment is used to help couples who cannot conceive a child naturally. It can also be used to help couples whose children are at risk from a serious genetic disorder.

Pre-implantation genetic diagnosis

Bob and Sally want children, but Bob has the allele for Huntington's disease. Sally has become pregnant twice. Tests showed that both the fetuses had the Huntington's allele and the pregnancies were terminated. Bob and Sally were keen to have a child, so their doctor suggested that they should use **pre-implantation genetic diagnosis (PGD)**. Sally's treatment is explained in the flow chart. The first use of PGD to choose embryos was in the UK in 1989. At the moment, PGD is only allowed for families with particular inherited conditions.

New technology – new decisions

New technologies like PGD often give us new decisions to make. In the UK, Parliament makes laws to control research and technologies to do with genes. Scientists are not free to do whatever research they may wish to do. From time to time Parliament has to update the law.

But Parliament can't make decisions case by case. So the Government has set up groups of people to decide which cases are within the law on reproduction. One of these groups **regulates** UK fertility clinics and all UK research involving embryos..

The regulatory body interprets the laws we already have about genetic technologies. It also takes into account public opinion, as well as practical and ethical considerations. One of its jobs is to decide when PGD can be used.

If embryos are tested before implantation, no babies need to be born with the disease. The embryos with faulty genes are destroyed. However, some people are concerned about this.

Questions

1 Explain why everyone who has PGD has to use IVF treatment in order to conceive.

2 Make a list of viewpoints for and against embryo selection.
   a Which viewpoint do you agree with? Give your reasons.
   b Which viewpoint do you disagree with? Give your reasons.

Cloning – science fiction or science fact?

Cloning: a natural process

Many living things only need one individual to reproduce. This is called **asexual reproduction**. Single-celled organisms like the bacteria in the picture use asexual reproduction.

The bacterium divides to form two new cells. The two cells' genes are identical to each other's. We call genetically identical organisms **clones**. The only variation between them will be caused by differences in their environment.

Asexual reproduction

Larger plants and animals have different types of cells for different jobs. As an embryo grows, cells become **specialised**. Some examples are blood cells, muscle cells, and nerve cells.

Plants keep some unspecialised cells all their lives. These cells can become anything that the plant may need. For example, they can make new stems and leaves if the plant is cut down. These cells can also grow whole new plants. So they can be used for asexual reproduction.

Some simple animals, like the *Hydra* in the picture opposite, also use asexual reproduction. But cloning is very uncommon in animals.

Sexual reproduction

Most animals and plants use **sexual reproduction**. The new offspring have two parents so they are not clones. But clones are sometimes produced – we call them identical twins.

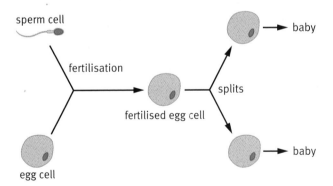

Identical twins have the same genes. But their genes came from both parents. So they are clones of each other, but not of either parent.

Cloning Dolly

Scientists can also clone adult animals artificially but this is much more difficult. Dolly the sheep was the first cloned sheep to be born.

- The nucleus is taken from an unfertilised sheep egg cell.
- The nucleus is taken out of a body cell from a different sheep.

Find out about

- ✔ asexual reproduction
- ✔ cloning and stem cells

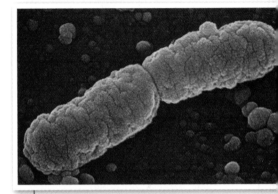

A bacterium cell grows and then splits into two new cells.
(Mag: × 7500 approx.)

Hydra.

Key words

- ✔ pre-implantation genetic diagnosis (PGD)
- ✔ regulates
- ✔ asexual reproduction
- ✔ specialised
- ✔ sexual reproduction
- ✔ clones

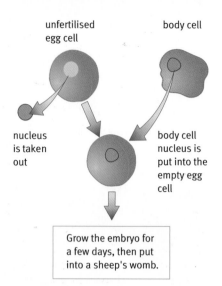

unfertilised egg cell

body cell

nucleus is taken out

body cell nucleus is put into the empty egg cell

Grow the embryo for a few days, then put into a sheep's womb.

- This body cell nucleus is put into the empty egg cell.
- The cell grows to produce a new animal. Its genes will be the same as those of the animal that donated the nucleus. So it will be a clone of that animal.

Dolly died in 2003, aged 6. The average lifespan for a sheep is 12–14 years. Perhaps Dolly's illness had nothing to do with her being cloned. She might have died early anyway. One case is not enough evidence to decide either way.

Many other cloned animals have suffered unusual illnesses. So scientists think that more research needs to be done before cloned mammals will grow into healthy adults.

Cloning humans

In the future it may be possible to clone humans. But most scientists don't want to clone adult human beings. However, some scientists do want to clone human cells. They think that some cloned cells could be used to treat diseases. The useful cells are called **stem cells**.

What are stem cells?

Stem cells are **unspecialised** cells. All the cells in an early embryo are stem cells. These embryonic stem cells can grow into any type of cell in the human body. Stem cells can be taken from embryos that are a few days old. Researchers use human embryos that are left over from fertility treatment.

Adults also have stem cells in many tissues, for example, in their bone marrow, brain, and heart. These unspecialised cells can develop into many, but not all, types of cells. Bone marrow stem cells are already used in transplants to treat patients with leukaemia.

Scientists want to grow stem cells to make new cells to treat patients with some diseases. For example, new brain cells could be made for patients with Parkinson's disease. But these new cells would need to have the same genes as the person getting them as a treatment. When someone else's cells are used in a transplant they are rejected.

What's cloning got to do with this?

Cloning could be used to produce an embryo with the same genes as the patient. Stem cells from this embryo would have the same genes as the patient. So cells produced from the embryo would not be rejected by the patient's body. Cloning a patient's own adult stem cells would produce cells that could be used to treat their illness. The cells would not be rejected because they are the same as the patient's own cells.

Doctors are exploring these technologies. Success could benefit millions of people if it is made to work.

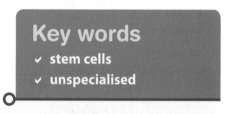

Key words
- ✔ stem cells
- ✔ unspecialised

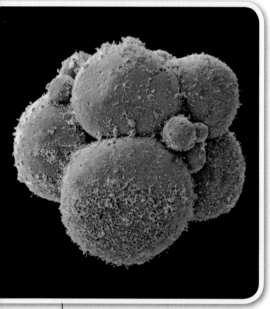

Cells from eight-cell embryos like this one can develop into any type of body cell. They start to become specialised when the embryo is five days old. (Mag: × 500 approx.)

Should human embryo cloning be allowed?

With some things there's no argument. Murder is just wrong – in the same way that lying and stealing are wrong. Killing an embryo at any age is as wrong as killing a child or an adult.

Research on embryos is legal up to 14 days. If something is 'legal' it can't be wrong.

James has Parkinson's disease. His brain cells do not communicate with each other properly. He cannot control his movements.

Whether it's right or not depends on how much good it does versus how much harm. If your best friend was paralysed in an accident, you wouldn't think it was wrong to sacrifice a five-day-old embryo made of 50 cells. Not if those cells could be used to make nerve tissue to repair your friend's damaged nerves.

An embryo is human so it has human rights. Its age doesn't make any difference. You can't experiment on a child or an adult.

Creating embryos for medical treatments is wrong. It's creating a life that is then destroyed. This lowers the value of life.

If research on cloning is allowed, it could lead to reproductive cloning. Once the technology to produce a human clone is developed, it will be difficult to stop someone using it to produce a cloned adult human.

Questions

1 How are stem cells different from other cells?

2 Explain why scientists think stem cells would be useful in treating Parkinson's disease.

3 For each of these cells, say whether or not your body would reject it:
   a bone marrow from your identical twin
   b your own skin cells
   c a cloned embryo stem cell.

4 For embryo cloning to make stem cells:
   a describe one viewpoint in favour
   b describe two different viewpoints against.

5 People often make speculations when they are arguing for or against something. This is something they think will happen, but may not have evidence for. Write down a viewpoint on human cloning that is a speculation.

Science Explanations

In this module, you will learn about inheritance, that genes are the units of inheritance, the relationship between genes and the environment, and that sexual reproduction is a source of variation.

You should know:

- that genes are sections of DNA and form part of chromosomes that are found in cells' nuclei and instruct cells to make proteins
- how single genes can determine some human characteristics, such as dimples; several genes working together determine many characteristics, such as eye colour; features such as scars are determined by the environment, and other characteristics, such as weight, are determined by both genes and the environment
- that a pair of chromosomes carries the same gene in the same place and alleles are different versions of a gene
- about dominant and recessive alleles and why the recessive gene in a heterozygous pair will not show its characteristics
- that a person's genotype is the genes they have, whilst their phenotype describes their characteristics
- how sex cells contain one chromosome from each pair and during sexual reproduction genes from both parents come together
- why offspring are similar to both parents but are not the same because they inherit a different combination of genes
- how a gene on the Y chromosome determines maleness
- that genetic diagrams are used to show inheritance
- how the symptoms of Huntington's disease appear later in life and include clumsiness, tremors, memory loss, and mood changes and that it is caused by a faulty dominant allele
- how a faulty recessive allele causes cystic fibrosis, a condition where cells produce thick mucus that causes chest infections and difficulties breathing and digesting food
- how genetic testing is used to screen adults, children, and embryos for faulty alleles
- how the information from genetic testing is used to make decisions and why this has implications
- that clones are organisms with identical genes
- about unspecialised cells, called stem cells, that can develop into other types of cell and how they can be used to treat some illnesses.

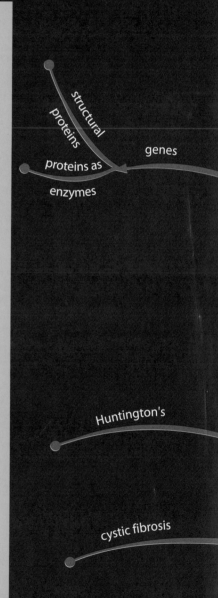

structural proteins

proteins as enzymes

genes

Huntington's

cystic fibrosis

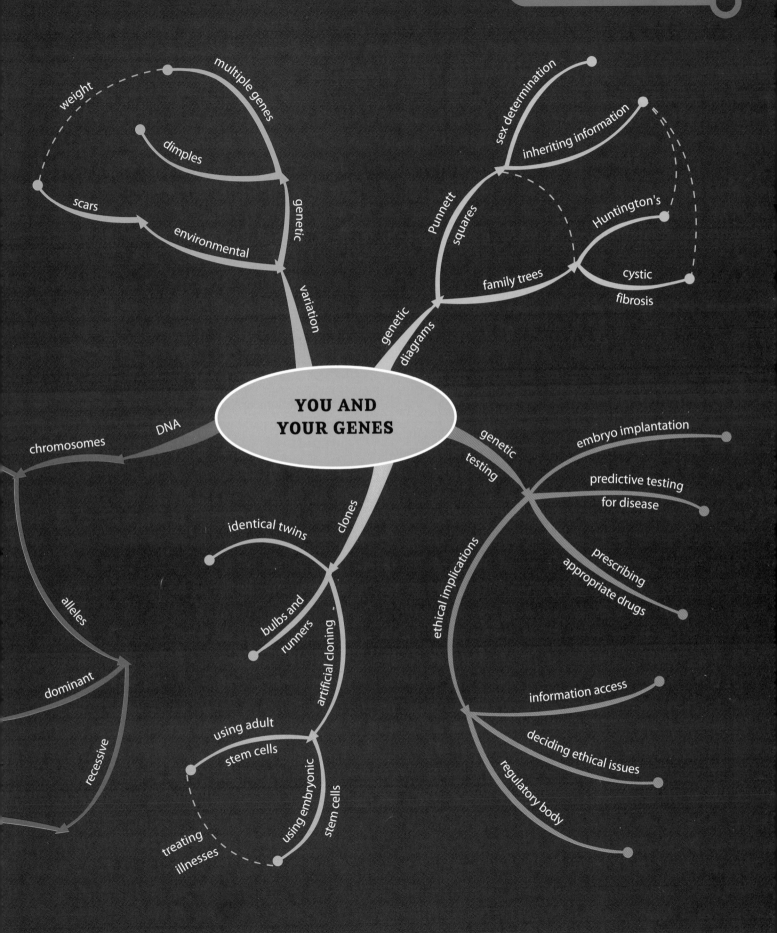

Ideas about Science

The application of science and technology has many implications for society. Ethical issues are often raised by science. The scientific approach cannot always answer these questions and society as a whole needs to discuss these issues and reach a collective decision.

Often the development and application of science is subject to regulation. You will need to be able to discuss examples of when this happens, for example:

• the role of the regulatory body for UK embryo research

• making decisions about genetic testing on adults and embryos prior to selection for implantation.

Some questions cannot be answered by science, for example, those involving values. You will need to be able to recognise questions that can be answered by using a scientific approach from those that cannot, such as:

• is it right to test embryos for genetic diseases prior to implantation?

• should genetic information about individuals be made available to the police or insurance companies?

Some forms of scientific work have ethical implications that some people will agree with and others will not. When an ethical issue is involved, you need to be able to:

• state clearly what the issue is

• summarise the different views that people might hold.

When discussing ethical issues, common arguments are that:

• the right decision is the one that leads to the best outcome for the majority of the people involved

• certain actions are right or wrong whatever the consequences and wrong actions can never be justified.

You will need to be able to:

• identify examples based on both of these statements

• develop ideas based on both of these statements.

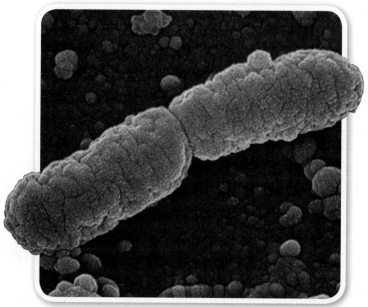

Review Questions

1 Cystic fibrosis is a genetic disorder.

a Choose the **two** words that describe the allele that causes cystic fibrosis:

faulty   normal

    dominant   recessive

b The family tree shows the inheritance of cystic fibrosis.

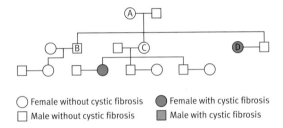

○ Female without cystic fibrosis   ● Female with cystic fibrosis
□ Male without cystic fibrosis   ■ Male with cystic fibrosis

i Which person, **A**, **B**, **C**, or **D**, is a female who has inherited two faulty cystic fibrosis alleles?

ii Which people from **A**, **B**, **C**, and **D**, are carriers?

iii Person **B** has a daughter. We cannot tell from the family tree if the daughter is a carrier. Explain why.

2 Science can show how genetic testing can be carried out. It cannot explain whether it should be carried out.

a Describe some implications of carrying out genetic testing on human beings.

b Explain the ethical issues involved in genetic testing and the different views that might be held.

3 Clones can be produced naturally.

a Which of the examples below are natural clones?

  A Two plants made by asexual reproduction from the same parent.

  B Two bacteria produced from one bacterium.

  C Identical twins.

  D Two sperm cells from the same man.

b Clones can look different.

Which factors can cause clones to look different? Choose the correct answer.

genetic factors only

environmental factors only

both genetic and environmental factors

neither genetic nor environmental factors

C1 Air quality

Why study air quality?

We breathe air every second of our lives. If it contains any pollutants they go into our lungs. Poor air quality can affect people's health. Chemicals that harm the quality of the air are called atmospheric pollutants. To improve air quality we need to understand how atmospheric pollutants are made.

What you already know

- The differences between solids, liquids, and gases.

- A mixture is made of two or more chemicals mixed together but not chemically combined.

- During a chemical change a new product is formed, with properties that are different from the reactants.

- Coal and natural gas are fossil fuels formed from the remains of living things.

- Elements are made up of just one type of atom.

- Each element is represented by a symbol (eg carbon = C).

- Compounds are made of two or more elements that are chemically combined.

- Compounds are each represented by a formula (eg water = H_2O).

- Data is used to provide evidence for scientific explanations.

Find out about

- the difference between 'poor-quality' and 'good-quality' air

- where the chemicals that harm air quality come from

- what can be done to improve air quality

- how scientists collect and use data on air quality

- how scientists investigate links between air quality and certain illnesses.

The Science

Most atmospheric pollutants are made by burning fossil fuels. When a fuel burns, the chemicals in the fuel combine with oxygen from the air. Some of the new chemicals that form are atmospheric pollutants. Burning is a chemical reaction. Knowing about chemical reactions helps people understand better what needs to be done to improve air quality.

Ideas about Science

Scientists who are trying to improve air quality measure the amounts of pollutants in the air. They must make sure their data is as accurate as possible. Some scientists use their data to see if they can find a link between air quality and health problems. New technologies to help reduce air pollution are always being developed, but only some are actually used. Deciding which ones to use can depend on cost, and on the most important needs at the time.

A | The air

Find out about

- the gases that make up air
- the atmosphere that surrounds Earth
- how other gases may be added to the atmosphere by human activity or natural processes

What do you know about air?

The air

Air is all around us. You cannot see the air but if you wave your hand you can feel it.

You may think that a can of fizzy drink is empty once you have drunk it, but look what happens if all the air is then removed from inside the can. The can collapses.

Air is made up of small **molecules** with large spaces in between. Molecules are groups of atoms joined together.

If you remove the air from inside a can, it collapses.

Key words

- molecule
- atmosphere
- mixture
- particulates

The atmosphere

An astronaut in space needs an air supply in order to breathe. A mountaineer climbing Everest feels that the air is becoming 'thinner' as he climbs. Where does our air stop?

The **atmosphere** is the layer of gases that surrounds the Earth. It is about 15 km thick. That sounds a lot but the diameter of the Earth is over 12 000 km. The atmosphere is like a very thin skin around the Earth.

The Earth from space. White clouds of water vapour can be seen in the atmosphere.

What gases are found in the air?

Air contains the gas oxygen. Oxygen is the gas we need to breathe but air is not just made of oxygen. Air is a **mixture** of oxygen and nitrogen, with a small amount of argon plus tiny amounts of carbon dioxide and water vapour.

The tiny amount of carbon dioxide in the air is what has helped keep the Earth warm enough to support life.

However, the air contains other gases as well. Human activity has released a whole variety of different gases into the atmosphere. Many of these gases affect the quality of the air we breathe. Unfortunately, gases released in one part of the world will gradually spread through the atmosphere and can affect the air quality of people many miles away.

Some gases are naturally released into the atmosphere by volcanoes. These gases include sulfur dioxide, carbon dioxide, carbon monoxide, nitrogen dioxide, and water vapour. They also produce **particulates** in the form of smoke and ash. The particulates in volcanic ash or smoke are tiny specks of solids. They are small enough to stay suspended in the air.

oxygen molecule containing two oxygen atoms

nitrogen molecule containing two nitrogen atoms

Nitrogen and oxygen make up 99% of the air.

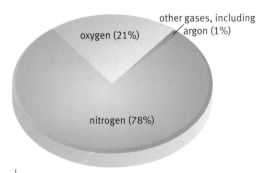

other gases, including argon (1%)

oxygen (21%)

nitrogen (78%)

This pie chart shows the percentages of the main gases in clean air.

This geologist is working upwind of a volcano. She is wearing a gas mask to protect her from breathing in sulfur dioxide gas.

Questions

1 Explain why a can does not collapse when the drink has been finished.

2 Oxygen, carbon dioxide and water vapour are three gases found naturally in the atmosphere. Explain why each of these is important.

3 In 2010 a volcano in Iceland erupted, releasing huge quantities of volcanic ash and particulates. Explain why it was not only local people in Iceland who were affected by this eruption.

Find out about

- ✔ the composition of gases in the Earth's early atmosphere
- ✔ evidence for an increase in oxygen in the atmosphere
- ✔ how scientists disagree about this evidence

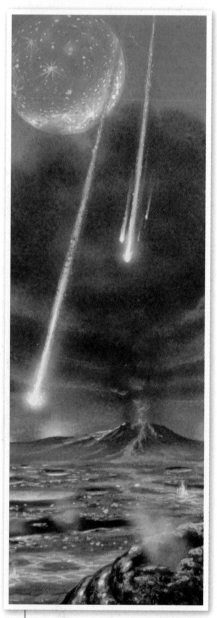

An artist's impression of early Earth.

Early Earth

The early Earth was a violent place, constantly bombarded by meteors and covered with active volcanoes. The atmosphere consisted mainly of carbon dioxide and water vapour. These gases probably came from volcanoes, which belched huge quantities of carbon dioxide and water vapour, and also some nitrogen and methane, into the atmosphere. If it had stayed that way, Earth would have been as inhospitable as our neighbouring planet Venus.

The temperature of the surface of Venus is almost 500 °C. Its atmosphere is largely carbon dioxide and its clouds are formed of sulfuric acid.

On Earth, temperatures began to cool. The water vapour gradually **condensed** to form the oceans. Some of the carbon dioxide began to dissolve in the oceans and later became incorporated into **sedimentary rocks**.

Exactly what happened next is less certain. Scientists are still investigating when and how the Earth's atmosphere changed from being mostly carbon dioxide to containing oxygen and just a little carbon dioxide.

How can scientists find out about the development of Earth's atmosphere?

Of course, scientists can't directly measure the composition of Earth's very early atmosphere. They have to use indirect evidence instead.

For example, the chemical composition of rocks can give scientists clues about the state of the atmosphere when the rocks were formed. So if they know the age of the rocks they find out something about the composition of the atmosphere at that time.

Scientists also look at fossil evidence of early life. The evidence suggests that oxygen levels first increased due to early plant life. These early plants used up carbon dioxide during **photosynthesis** and released oxygen. So as the oxygen level rose, the carbon dioxide level fell. Much of the carbon dioxide was removed from the atmosphere for a long time because it was trapped underground as the coal and oil that we now use as fuels.

Iron pyrite is made up of iron sulfide, which only forms if there is *no* oxygen present.

Air bubbles trapped in ice cores drilled in Antarctica allow scientists to analyse the composition of the air from hundreds of thousands of years ago. However, the Earth is about 4.5 billion years old. Ice cores don't go back far enough in time to provide evidence of the earliest atmosphere.

The details of exactly when and how fast these changes in the atmosphere happened are still a matter of scientific debate, as scientists discuss and evaluate each others' ideas and explanations.

Scientists may interpret available data in different ways. An expert in plants might interpret fossil data differently from an expert in rocks. This may lead them to come up with different scientific explanations for when and how the atmosphere changed.

Even if scientists agree on an explanation, new information may then be found that disagrees with this explanation. For example, in 2010 some American scientists reported that they had found fossils of primitive animals with shells that lived in ocean reefs 650 million years ago. These fossils are 50–100 million years older than any other known fossils of hard-bodied animals. New discoveries like this mean that scientists have to rethink their ideas and come up with new explanations to account for fresh data.

Red iron oxide rocks only form if there is oxygen present.

Questions

1 Rocks made of iron oxide have been dated to two billion years ago. What does this tell you about when oxygen first appeared in Earth's atmosphere?

2 No samples of iron pyrite have been dated as less than two billion years old. What does this tell you about when oxygen first appeared in the Earth's atmosphere?

3 What discovery could change your mind about these conclusions?

Key words
- ✓ **condensed**
- ✓ **photosynthesis**
- ✓ **sedimentary rock**

Find out about

- ✔ the most important air pollutants
- ✔ the problems pollutants cause
- ✔ what can influence air quality in different locations

Every time you are driven somewhere by car, or when you switch on the lights at home, new gases are made. They are released by the car, or from the power station where electricity is generated. Some of these chemicals are harmful and are called air **pollutants**. These pollutants can harm us directly by affecting our health, or harm us indirectly by affecting our environment.

Air pollutants spread through the atmosphere and change its composition. The biggest recent change has been in levels of carbon dioxide, which have been increasing for over a century.

The table lists the main chemicals that are released from power stations and vehicles.

The clouds coming from the cooling towers may look like pollutants, but are just harmless water vapour. There may be invisible pollutants coming out of the tall chimney.

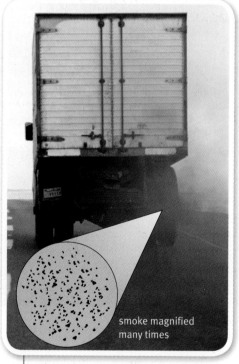

smoke magnified many times

Smoke is a pollutant that can be easily seen. It contains microscopic particles of carbon. Some of these are just 10 micrometres (10 millionths of a metre) in size. These are called PM10 particles. Although they are very small, they are very much bigger than atoms or molecules. Each particle contains billions of carbon atoms.

sulfur dioxide SO_2		Reacts with water and oxygen to produce acid rain. This can damage buildings and harm trees and plants.
carbon monoxide CO		A poisonous gas. Changes the amount of oxygen in the blood. This can make people's existing heart conditions worse.
carbon dioxide CO_2		Dissolves in rain water and sea water. Used by plants in photosynthesis. Excess levels of CO_2 can give rise to global warming.
nitrogen monoxide NO		Reacts in the atmosphere to form nitrogen dioxide.
nitrogen dioxide NO_2		Reacts with water and oxygen to produce acid rain. Can cause breathing problems and can make asthma worse.
particulates (tiny bits of solid suspended in the air)		Deposited on surfaces, making them dirty. Can be breathed into the lungs and can make asthma and lung infections worse.
water H_2O		Harmless. Not a pollutant.

How can you find out about air quality?

Some people suffer from asthma or hayfever. They may be able to feel when the air quality is poor. But most people do not know whether the air quality is good or bad.

Automatic instruments collect air samples and measure the concentrations of a range of pollutants. The data is recorded automatically. Much of the data is regularly displayed in real time on websites available to the public. Newspapers and TV stations summarise the data in reports, which may give the day's overall air quality on a number scale or describe it as low, medium, or high quality.

Does it matter where you live?

Is the air quality the same all over the country? Some people live in cities. Other people live in the countryside. Will they all have air of the same quality to breathe?

The bar chart shows the concentration of NO_2 on the same day in three different places. The concentrations are clearly different. The concentration of NO_2 depends a lot on the level of human activity in the area. The amount of road traffic has a big effect.

Mace Head, in Ireland, has very pure air when the wind blows in from across the Atlantic Ocean. Scientists use it as a baseline to see what air would be like without the effects of human activities.

Most of us live in environments where the air quality is much poorer than at Mace Head.

Measuring the concentration of a pollutant

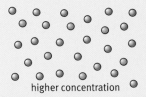

A low concentration of pollutants. There are very few pollutant molecules in a certain volume of air. This is an indication of good air quality.

lower concentration

A high concentration of pollutants. There is a large number of pollutant molecules in a certain volume of air. This shows that the air quality is poor.

higher concentration

○ molecules of pollutant
○ other molecules in air

Concentration is the amount of pollutant in a certain volume of air.

(Note: the air molecules are normally much more spread out than shown in the diagrams.)

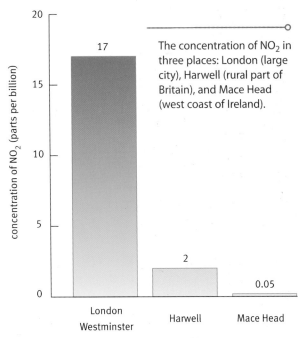

The concentration of NO_2 in three places: London (large city), Harwell (rural part of Britain), and Mace Head (west coast of Ireland).

An air-pollution monitoring station in a busy London street. It has instruments to measure particulate matter, carbon monoxide, nitrogen dioxide, and sulfur dioxide levels.

Nitrogen dioxide in London

Nitrogen dioxide levels increase when traffic is heavy. Exhaust **emissions** released from vehicles contain nitrogen dioxide. Can you see any patterns in the graph below that back this up?

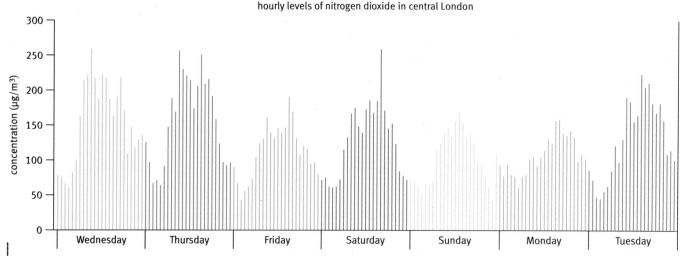

hourly levels of nitrogen dioxide in central London

Nitrogen dioxide levels in central London over a seven-day period at the beginning of January 2009.

What influences air quality?

The quality of the air where you live depends mostly on nearby sources of air pollutants, and the weather.

- Sources: vehicles, power stations, and industry are some of the main sources of air pollutants.
- Weather: pollutants are mixed up and carried around by the winds. Wind can move pollutants many miles and even carry them from one country to another.

Questions

1 Write down one problem that can be caused by each of these air pollutants:

  a SO_2 b NO_2 c particulates

2 A newspaper article on air quality included a photograph of white clouds coming out of a power station's cooling towers. Write a note to the paper explaining why the clouds are not polluting the atmosphere.

3 Look at the chart above showing levels of NO_2 in London. Suggest reasons for the pattern of readings for Wednesday.

4 The weather moves air pollutants from one place to another. If we reduce emissions of air pollutants in our own town, we can still get pollution from other areas. Explain why it is still important to try to reduce emissions.

Crucial data

Quantities of air pollutants are measured using a network of monitoring stations all around the country. The information is useful to individual people, as it allows them to check the air quality in their area. It is also used by the government to check whether air pollution is reaching dangerous levels anywhere.

Scientists use data to help answer questions such as 'how do pollutants travel?' or 'how do air pollutants interact with other chemicals in the atmosphere?' Data can be used to check the scientists' proposed explanation.

For example, the data from a busy street showed that the concentration of nitrogen dioxide (NO_2) was much higher on one side than the other. One explanation was that it could be caused by patterns in air movements in the street.

More data were collected to test this explanation. The computer-generated picture below shows the concentrations of nitrogen dioxide at all levels in the street. You can see that the air currents are actually circulating. This is invisible to people in the street, but obvious when you use the data.

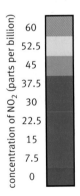
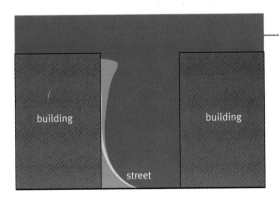

This computer-generated picture shows the concentration of the air pollutant NO_2 in a city street. The red area shows that invisible air currents have channelled the NO_2 onto just one side of the street. The tall buildings on either side make the street into a kind of 'canyon' and trap the NO_2 at street-level.

Making measurements

If you measure the concentration of nitrogen dioxide in a sample of air several times, you will probably get different results. This is because:

• you used the equipment differently
• there were differences in the equipment itself.

If you take just one reading, you cannot be sure it is very accurate. So it is better to take several measurements. Then you can use them to estimate the true value.

The true value is what the measurement should really be. The **accuracy** of a result is how close it is to the true value.

Key words
- ✔ accuracy
- ✔ outlier
- ✔ mean value
- ✔ best estimate
- ✔ range
- ✔ real difference

How can you make sure your data is accurate?

The table below shows what you should do to get a measurement of the level of nitrogen dioxide that is as accurate as possible.

The mean value is 19.1 ppb. This is the best estimate of the concentration of nitrogen dioxide in the sample of air. You cannot be absolutely sure that it is the true value. But you can be sure that:

- the true value is within the range 18.8 – 19.4 ppb
- the best estimate of the true value is 19.1 ppb.

If you had taken only one measurement, you wouldn't have been sure it was accurate. If the range had been narrower, say 19.0 – 19.3 ppb, you would have been even more confident about your best estimate of the true value.

What you do	Data	Describing what you do
Take several measurements from the same air sample. Not all the measurements will be the same.	Concentration of NO_2 in parts per billion (ppb) 18.8, 19.1, 18.9, 19.4, 19.0, 19.2, 19.1, 19.0, 18.3, 19.3	The measurements (10 in this case) are called the data set.
Plot the results on a number line. This shows that the 18.3 ppb measurement is very different from the others. Decide whether to ignore this reading.	19.6 ⊢ 19.5 ⊢ 19.4 ⊢ × 19.3 ⊢ × 19.2 ⊢ × 19.1 ⊢ × × 19.0 ⊢ × × 18.9 ⊢ × 18.8 ⊢ × 18.7 ⊢ 18.6 ⊢ 18.5 ⊢ 18.4 ⊢ 18.3 ⊢ ⊗ —— this result is an outlier	A result that is very different from the others is called an **outlier**. If you can think of a reason why this result is so different (eg, you made a mistake when you took this measurement), you should ignore it.
Add the other nine results together. Divide the total by 9. The answer is 19.1 ppb of NO_2.	Total of nine readings = 171.8 $\frac{171.8}{9} = 19.1$ ppb	19.1 is called the **mean value** of the nine measurements.
You can use the mean value rather than any of the nine measurements.	The best estimate for the concentration of NO_2 is 19.1 ppb	The mean value is used as the **best estimate** of the true value.
When you write down the mean value you also record: • the lowest, 18.8 ppb, • and the highest,19.4 ppb, measurements.	The range is 18.8 – 19.4 ppb	18.8 – 19.4 ppb is called the **range** of the measurements.

Comparing NO$_2$ concentrations

The graph on the right shows the mean and range for the concentration of nitrogen dioxide in three different places.

- Compare London and York. The means are different but the ranges overlap.
- The range for London overlaps the range for York. So the true value for London could be the same as the true value for York. You cannot be confident that their NO$_2$ concentrations are different.
- Compare London and Harwell. The means are different and the ranges do not overlap.
- You can be very confident that there is a real difference between the NO$_2$ concentrations in London and Harwell.

When you compare data, do not just look at the means. To make sure that there is a **real difference**, check that the ranges do not overlap.

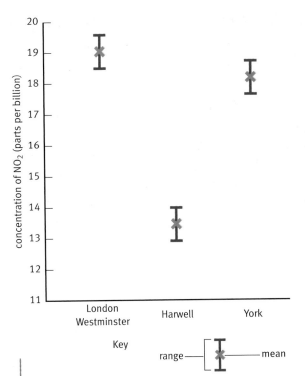

Key

range ⊢✕⊣ mean

NO$_2$ concentrations in air from three places in England. All the measurements were made at the same time of day.

Questions

1 Jess measured the NO$_2$ concentration in the middle of a town. She took six readings: 22 ppb, 20 ppb, 18 ppb, 24 ppb, 21 ppb, 23 ppb. Jess used new equipment and was careful taking her measurements.
   a Calculate the mean value of the measurements.
   b Write down the best estimate and the range for the NO$_2$ concentration in this sample of air.

2 Look at the graph above. Does it show that there is a real difference in NO$_2$ levels between Harwell and York? Explain your answer.

3 Repeat measurements on an air sample produced these results for the NO$_2$ concentration:

   Reading 1 – 39.4 ppb   Reading 2 – 45.8 ppb
   Reading 3 – 42.3 ppb   Reading 4 – 38.7 ppb
   Reading 5 – 39.7 ppb   Reading 6 – 32.7 ppb

There had been some problems with the equipment that day.
   a Plot these six readings on a number line.
   b Work out the mean NO$_2$ concentration and range for this sample.
   c Another sample was taken from a second place in the same town. The mean NO$_2$ concentration for this sample was found to be 44.1 ppb. Can you say with confidence that the second location had a higher NO$_2$ concentration than the first? Explain your answer.

4 A scientist took one measurement of NO$_2$ in an air sample. Explain why this would not give you much confidence in the accuracy of the result.

Find out about

✓ what fuels are made from
✓ the products that result from burning fuels
✓ how burning fuels can make atmospheric pollutants

Many atmospheric pollutants are made by the burning of fossil fuels. This happens in power stations and in the engines of vehicles.

What happens when fuel burns in a power station?

Most power stations are fuelled by either coal or natural gas. Fuel and air go into the furnace and waste chemicals come out of the chimney. Use the diagram below to compare what goes in with what comes out of each type of power station.

Any change that forms a new chemical is called a **chemical change** or a **chemical reaction**.

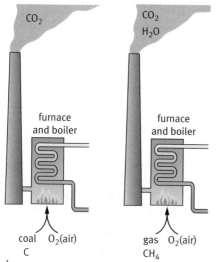

The chemicals going into and coming out of power station furnaces.

Coal-fired power station

Coal is mainly made up of carbon atoms. Oxygen molecules from the air are needed for coal to burn. The main product that comes out of the chimney of a coal-fired power station is carbon dioxide (CO_2). It must have been formed by the oxygen molecules (O_2) separating into oxygen **atoms**. These oxygen atoms then combine with the carbon atoms to make carbon dioxide. Reactions where a chemical joins with oxygen are called oxidation reactions.

Gas-fired power station

Natural gas is mainly methane (CH_4). Methane is a **hydrocarbon**. It is made of carbon and hydrogen atoms.

The main products from the burning of natural gas are CO_2 and H_2O.

These must have been formed by:

- carbon atoms and hydrogen atoms in CH_4 separating
- then carbon atoms combining with oxygen atoms to form CO_2
- and hydrogen atoms combining with oxygen atoms to form H_2O.

A hydrocarbon molecule. Natural gas, petrol, diesel, and fuel oil are all mainly made up of hydrocarbon molecules.

Other products from burning fuel

Burning coal and gas can also produce smaller amounts of these air pollutants:

- particulates – small pieces of unburned carbon
- carbon monoxide (CO) – formed when carbon burns in a limited supply of oxygen
- sulfur dioxide (SO_2) – formed if the fuel contains some sulfur atoms.
- nitrogen monoxide (NO) – formed when some of the nitrogen in the air reacts with oxygen at the high temperatures in the furnace

Nitrogen monoxide (NO) can then react with oxygen in the air to form nitrogen dioxide (NO_2). This is an oxidation reaction. Together, NO and NO_2 are referred to as nitrogen oxides.

Key words

✓ chemical change/reaction
✓ atom
✓ hydrocarbon

What happens when fuel burns in a car engine?

Vehicle engines burn petrol or diesel. These are also made up of hydrocarbon molecules. Use the diagram to compare what goes into a car engine with what comes out.

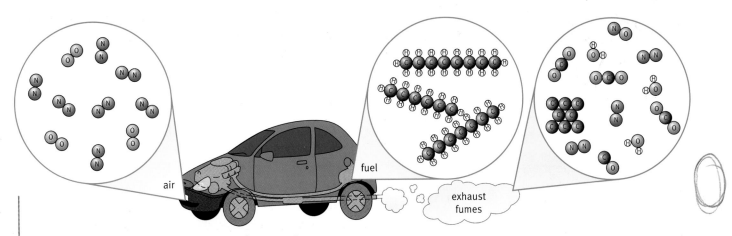

The chemicals going into and coming out of a car engine.

The overall change can be summarised as:

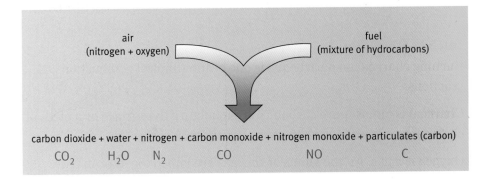

air (nitrogen + oxygen) → fuel (mixture of hydrocarbons)

carbon dioxide + water + nitrogen + carbon monoxide + nitrogen monoxide + particulates (carbon)
CO_2 H_2O N_2 CO   NO   C

Check that you know which of the pictures in the three circles represents each of the chemicals mentioned in the summary. You can use these pictures to work out the chemical changes happening in the engine, and therefore how the pollutants are formed. For example, nitrogen monoxide (NO) must have been formed from nitrogen (N_2) and oxygen (O_2) in the air. These must have first split apart into atoms and then combined to make NO.

Questions

1 List the air pollutants that can be released from a coal-burning power station.

2 List the air pollutants released from a car engine when it burns fuel.

3 Use ideas about atoms separating and then joining together in different ways to explain how:
   a H_2O forms when methane gas (CH_4) burns in a power station.
   b CO forms when coal (C) burns in a power station.
   c CO_2 forms when petrol burns in a car.

Find out about

✓ **how atoms are rearranged during combustion reactions**
✓ **different ways of representing chemical changes**

Reactions where a chemical joins with oxygen are called **oxidation** reactions.

Some chemicals can react rapidly with oxygen to release energy and possibly light. This type of oxidation reaction is called **combustion** or burning.

Fuel has escaped during this racing car crash. An uncontrolled combustion reaction is happening. The fuel and air mixture has been heated by either a spark or the hot engine.

Burning charcoal

Burning charcoal on a barbeque is one of the simplest combustion reactions.

Charcoal is almost pure carbon. You can picture the surface of a piece of charcoal as a layer of carbon atoms tightly packed together.

Oxygen is a gas. All the atoms of this gas are joined together in pairs (O_2). These are molecules of oxygen.

During a combustion reaction, the atoms of carbon and oxygen are rearranged.

It will help you to understand this reaction if you can picture what happens to the atoms and molecules involved.

Fuel burns more rapidly in pure oxygen than in air. Oxygen obtained from the atmosphere is used in this oxy-fuel welding torch.

Air contains oxygen gas. One molecule of oxygen is two oxygen atoms joined together ⚪⚪. Oxygen molecules split and react with carbon atoms in the charcoal. This forms carbon dioxide gas ⚪●⚪.

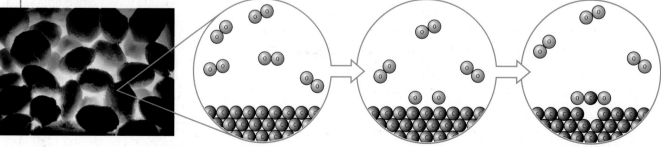

Describing combustion reactions

You can use pictures to describe the chemical change that happens when carbon dioxide burns.

 + ⟶

The chemicals before the arrow are the ones that react together. We call them **reactants**.

The chemicals after the arrow are the new chemicals that are made. We call them **products**.

It would be time consuming if you always had to draw pictures to describe chemical reactions. So scientists use equations to summarise the pictures.

The combustion of charcoal can be summarised in this **word equation**:

carbon + oxygen ⟶ carbon dioxide

If you want more detail, you can write the **chemical equation** to show the atoms that make up each of the chemicals involved. This uses symbols for each chemical. These are called **chemical formulae**.

This is the chemical equation for the combustion of charcoal.

$$C + O_2 \longrightarrow CO_2$$

The chemical equation is a more useful description of the reaction than the word equation. It tells you how many atoms and molecules are involved and what happens to each atom.

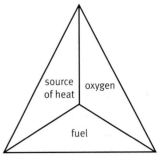

Three things are needed for a fire, or combustion reaction:
- a **fuel**, mixed with
- **oxygen** (air), and a
- source of **heat**, to raise the temperature of the mixture.

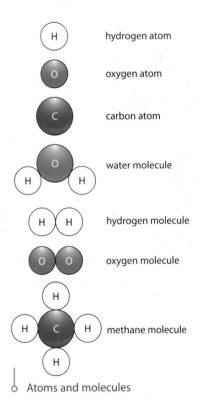

Atoms and molecules

Questions

1 How is a welding torch designed to produce a flame hot enough to melt and join metal?

2 What are the reactants and what are the products in each of the following chemical changes:
   a carbon combines with oxygen to form carbon dioxide
   b a hydrocarbon in petrol burns in oxygen to form carbon dioxide and water.

3 You and your cousin are having a barbecue. Your cousin asks you what happens to the charcoal when it burns. Write down what you would say. Include the words: atom, molecule, combustion, reactants, products, chemical change.

4 Draw pictures to represent these chemical changes:
   a hydrogen burning in oxygen to form water
   b methane burning in oxygen to form water and carbon dioxide.

Key words

- ✓ **oxidation**
- ✓ **combustion**
- ✓ **reactants**
- ✓ **products**
- ✓ **chemical formula**
- ✓ **word equation**
- ✓ **chemical equation**

Find out about

- ✓ **what happens to atoms during chemical reactions**
- ✓ **how the properties of reactants and products are different**

When you have had a bonfire, some of the atoms that made up the rubbish are in the ashes left on the ground. The others are in the products released into the air.

Look at the picture below. How many atoms of hydrogen (H) are there before and after the reaction? Count the atoms of oxygen (O) before and after. What does this show?

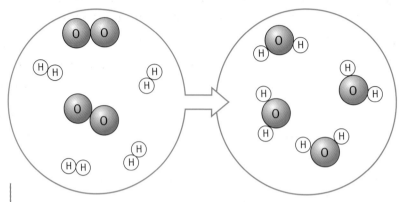

The reaction of hydrogen and oxygen to form water.

Conservation of atoms

All the atoms present at the beginning of a chemical reaction are still there at the end. No atoms are destroyed and no new atoms are formed. The atoms are conserved. They rearrange to form new chemicals but they are still there. This is called **conservation of atoms**.

For example, when a car engine burns fuel, the atoms in the petrol or diesel are not destroyed. They rearrange to form the new chemicals found in the exhaust gases.

Look again at the picture of hydrogen reacting with oxygen to form water. *Two* molecules of hydrogen react with just *one* molecule of oxygen. This produces *two* molecules of water. We can represent this change by:

Notice that there are the same numbers of each kind of atom on each side of the equation. All the atoms that are in the reactants end up in the products. The atoms are conserved.

All atoms have mass. Because the atoms are conserved, the mass of the reactants is the same as the mass of the products. This is called **conservation of mass**.

Properties of reactants and products

The **properties** of a chemical are what make it different from other chemicals.

For example, some chemicals are solids, some are liquids, and some are gases at normal temperatures. Some are coloured, some burn easily, some smell, some react with metals, some dissolve in water, and so on. Each chemical has its own set of properties.

The table compares the properties of the reactants and products of the reaction between sulfur and oxygen.

Chemical	Properties
sulfur (reactant)	yellow solid
oxygen (reactant)	colourless gas; no smell; supports life
sulfur dioxide (product)	colourless gas; sharp, choking smell; harmful to breathe; dissolves in water to form an acid

In any chemical reaction, all the atoms you start with are still there at the end. But they are combined in a different way. So the properties of the products are different from the properties of the reactants.

This is very important for air quality. You can have a piece of coal that is a harmless black stone.

But the coal may contain a small amount of sulfur. When it burns the sulfur will react with oxygen to form the gas sulfur dioxide.

The sulfur dioxide escapes into the atmosphere. It will harm the quality of our air. It will react with water and oxygen in the air to form acid rain. This is harmful to plants, animals, and buildings. The harmless piece of coal has produced a harmful gas.

The atoms and molecules involved in the burning of sulfur.

Key words

- ✓ **conservation of atoms**
- ✓ **conservation of mass**
- ✓ **properties**

Questions

1 You have learned that atoms are conserved during a chemical reaction. Work out how many molecules of CO_2 and H_2O will be produced when one molecule of methane (CH_4) is burnt. Draw a picture to show the atoms and molecules in the reaction. Work out how many molecules of O_2 will be used.

2 Water (H_2O) is made by reacting molecules of hydrogen and oxygen together. How are the properties of water different from the properties of the reactants it is made from?

3 Burning rubbish gets rid of it forever. Is this a true statement? Think about the atoms in the rubbish. Fully explain your answer.

Find out about

- how to look for links between air-quality data and the symptoms of an illness
- how pollen causes hayfever
- the link between asthma and air quality

Hayfever

Do you suffer from a runny nose, sneezing, and itchy eyes in the summer? This could be hayfever.

Hayfever got its name because people noticed that it happens in the summer. This is when grass is being cut to make hay. It is also the time when pollen from plants is at its highest.

To find out what causes hayfever it is important to first look at what **factors** are linked with hayfever. Factors are variables that may affect the outcome. In this case, hayfever is the **outcome.** Pollen is a factor that may affect hayfever.

Pollen traps collect pollen grains so that they can be counted using a microscope. This gives the 'pollen count'. Newspapers, radio, and television report the pollen count during the summer.

Is there a link between hayfever and pollen?

If an outcome increases (or decreases) as a particular factor increases, this is called a correlation. So, do more people suffer from hayfever when the pollen count increases?

Looking at thousands of people's medical records shows that hayfever is highest in the summer months. This is when most pollen is in the air. It is important to look at a randomly selected sample of medical records to collect data that is representative of the whole population.

This evidence shows that there is a correlation between pollen levels and hayfever symptoms. But does this mean that there is a causal link between pollen and hayfever – is pollen the **cause** of hayfever?

Pollen is released by plants and may travel many kilometres on the wind. Pollen grains are in the air that we breathe.

Pollen grains under the microscope. Different plants release different types of pollen. (Magnification approximately ×1360.)

Does pollen cause hayfever?

An increase in two things could be caused by a third factor that has not been measured. Or it could be a coincidence that the two things increase at the same time.

Think about ice cream. Most ice cream is sold in the summer months but nobody would say that ice cream causes hayfever. It may be just a coincidence. Or both increases may be caused by some other factor.

To claim that pollen causes hayfever you need some supporting evidence. You need to show that there is a correlation and you need to be able to explain *how* pollen causes hayfever.

Different types of pollen are released at different times of year. Some people have hayfever in months that correlate with particular types of pollen being released. This is strong extra evidence for the correlation between pollen and hayfever.

Skin-prick tests show that people who suffer from hayfever are allergic to pollen. Hayfever is an allergic reaction caused by pollen. So there is a **correlation** between hayfever and pollen, because pollen causes hayfever.

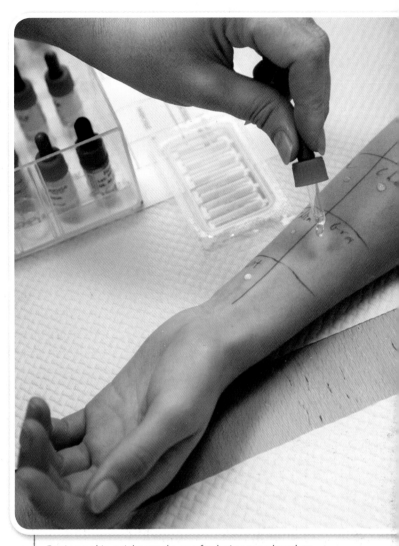

During a skin-prick test, drops of solution are placed on the skin. The skin beneath is pricked. If the patient is allergic to the substance in the solution (eg pollen), their skin will turn red and itchy.

Questions

1 Suggest why it is useful to report the levels of pollen in the air during the summer months.

2 Write a note to a friend explaining:
   a what is meant by 'there is a correlation between pollen count and hayfever symptons'
   b why you need to look at medical records of a large number of people to be sure there is a correlation
   c why a correlation between ice cream sales and hayfever does not mean that ice cream causes hayfever.

Key words
- ✓ **correlation**
- ✓ **cause**
- ✓ **factor**
- ✓ **outcome**

Asthma and air quality

Asthma is a common problem. During an asthma attack, a person's chest feels very tight. They find it difficult to breathe.

Many things can trigger asthma attacks in people who have asthma. These include:

- tree or grass pollen
- animal skin flakes
- dustmite droppings
- air pollution
- nuts, shellfish
- food additives
- dusty materials
- strong perfumes
- getting emotional
- stress
- exercise (especially in cold weather)
- colds and flu.

Inhalers are used to treat asthma attacks and to keep the condition under control. They contain medicines that help people's airways 'open up', allowing them to breathe more freely.

Causes of asthma

Medical evidence shows that asthma attacks are triggered by many different factors.

Nitrogen dioxide is an air pollutant that comes mainly from traffic exhausts. Large-scale studies have shown that if the concentration of nitrogen dioxide stays high for several days, there is an increased risk of people with asthma suffering from asthma attacks. This is a correlation.

People who have asthma have sensitive lungs. Air pollutants may irritate a person's lungs, particularly if their lungs are sensitive. If the level of nitrogen oxide pollution stays high for several days there is an increase in the number of asthma attacks. The data shows that there is a correlation between levels of pollution and asthma, but the data does not give clear evidence that nitrogen oxides cause asthma.

Exposure to nitrogen oxides increases the chance of an asthma attack but does not mean that all people with asthma will suffer an attack. Even so, the link between the factor (exposure to polluted air) and the outcome (an asthma attack) is still described as a correlation.

Studying asthma

The causes of asthma are not fully understood. There is evidence to show that some people are more likely to have asthma attacks because of their genes. Other evidence indicates that environmental factors, such as air pollution, are also involved.

Scientists who study asthma publish their findings at conferences and in journals. This makes it possible for other scientists to evaluate their claims critically. Reviewers look carefully at the way that the scientists describe their methods of investigation, at their presentation of the data, and at the way they interpret that data and come to conclusions.

It has been suggested, for example, that stress could be a factor that increases the chance of an asthma attack. Supporters of the stress theory suggest that stress changes the way the lining of the lungs reacts to irritants.

The stress theory is just one of several explanations for the worldwide increase in the number of people affected by asthma. The data is complex and hard to interpret. This means that different scientists come up with different conclusions about the causes of asthma.

Exposure of children to high levels of traffic pollution may lead to an increased chance of their developing asthma.

Questions

3 a Make a list of some of factors that might be a cause of asthma.

   b Explain why scientists are finding it hard to work out why more and more people are being affected by asthma.

4 Why is it important that scientists publish their data and explanations?

5 Why do scientists look to see if published results can be replicated by other scientists before deciding to accept a scientific claim?

Find out about

- how laws and regulations can help improve air quality
- how new technology can reduce harmful emissions from cars and power stations
- what we can all do to reduce air pollution

Reducing pollutants from cars

For individuals, the simplest way to improve air quality is to use the car less. A good alternative is to use public transport such as buses or trains, or walk or cycle. Fifty people travelling by bus use a lot less fuel than if they each travel in their own private car. Less fuel burnt means fewer pollutants released. By making it easier for people to use buses and trains, governments can help improve air quality.

Governments can also set legal limits on the amount of pollutants a vehicle is allowed to produce. In Britain, MOT tests include a vehicle emissions test, which means that cars with emissions above the legal limit are not allowed on the roads. **Regulations** state that vehicles over three years old must have an MOT test every year.

Efficient engines and catalytic converters

Engineers are continually working on improving the **efficiency** of car engines. A more efficient engine means that a car will burn less fuel to travel the same distance, reducing pollutants. This is good for car owners too because they do not need to buy so much fuel.

Even a very efficient engine will still produce air pollutants. This is why scientists have developed ways of removing the worst pollutants from the exhaust. All new cars have catalytic converters fitted to their exhaust systems. The waste gases pass through a metal honeycomb structure with a large surface area. The metal surface speeds up certain chemical reactions. A **catalytic converter** changes the pollutants carbon monoxide (CO) and nitrogen monoxide (NO) into less harmful gases.

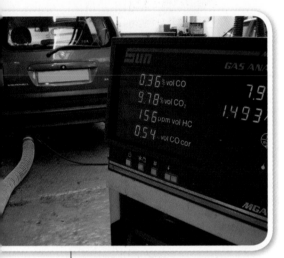

MOT exhaust emission analysis. Exhaust emissions are tested for carbon monoxide and unburnt fuel.

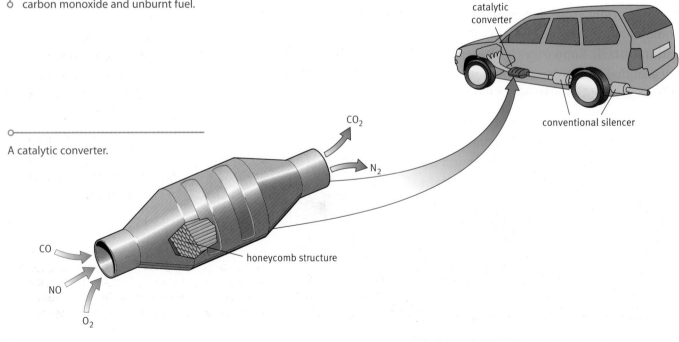

A catalytic converter.

catalytic converter

conventional silencer

CO_2

N_2

honeycomb structure

CO

NO

O_2

The chemical reactions that occur in a catalytic converter are:

carbon monoxide + oxygen $\longrightarrow$ carbon dioxide

nitrogen monoxide $\longrightarrow$ nitrogen + oxygen

Oxygen is added to carbon monoxide. This is an oxidation reaction.

Oxygen is removed from nitrogen monoxide. This is a **reduction** reaction.

Even with a catalytic converter, carbon dioxide is still released. Carbon dioxide levels are a concern due to the link with global warming. The only way of producing less carbon dioxide is to burn less fossil fuel.

Cleaner transport

Transport can be made cleaner by improving existing fuels or using other sources of power.

Diesel fuel contains small amounts of sulfur compounds. Low-sulfur fuels have had these compounds removed. This reduces the amount of sulfur dioxide in exhaust emissions.

Biofuels are a renewable alternative to fossil fuels for motor vehicles. They are made from plants such as sugar cane, corn, and oilseed rape. The plants absorb carbon dioxide as they grow. When the fuels are burnt, this carbon dioxide returns to the atmosphere. Producing more biofuel would need more land given over to growing plants for fuel instead of food. A small amount (2–3%) of biofuel can be mixed into petrol or diesel without changing engines or fuel stations.

Some vehicles are now powered using electricity stored in batteries, rather than fuels such as petrol, diesel, or biofuels. Electric vehicles do not make any waste gases when they are used. However, the electricity comes from power stations, which in many cases burn fossil fuels. Using electric vehicles in congested cities means the pollutants are not released in the city but elsewhere. In comparison with petrol vehicles, electric vehicles have a short range before needing to be recharged. Recharging can take several hours.

Key words

✓ efficiency
✓ catalytic converter
✓ reduction
✓ regulations

Questions

1 All new cars are fitted with a catalytic converter.
  a Which two pollutants are removed by catalytic converters?
  b For each pollutant, which less harmful gas is it changed into? What type of reaction is needed to do this?
  c Why is this not a perfect solution?

2 Hybrid cars use electricity, but also have a normal fuel engine for when it is needed.
  a Explain why using a hybrid car could reduce traffic pollution in a town.
  b Explain why using a hybrid car in electric mode can still result in pollutants reaching the atmosphere.

3 Write a letter to your local councillor, suggesting how people living in your area could reduce air pollution.

This hybrid car runs on both electricity and petrol. The electricity comes from power stations through the national grid. By choosing a car that can use electricity, drivers can reduce the amount of petrol they use.

Trees killed by acid rain in the Czech Republic. Sulfur dioxide is a waste gas produced by power stations. It reacts with water and oxygen to form acid rain.

Reducing pollutants from power stations

Individuals, governments, and industries all have a role to play in reducing air pollution from power stations.

Each individual can contribute towards a reduction in pollutants released from power stations by using less electricity. However, even if people use less electricity, it is still good to reduce the air pollution produced in making the electricity we do use.

Most power stations burn coal, natural gas, or fuel oil. When these fossil fuels are burnt, waste gases and particulates are produced. The main product is carbon dioxide. Fossil fuels also contain impurities such as sulfur. When sulfur burns, the pollutant sulfur dioxide is produced. This can go on to produce acid rain.

Sulfur dioxide

One way of reducing the amount of sulfur dioxide released into the atmosphere is to remove sulfur from fuels before they are burnt.

Natural gas and fuel oil can be refined to remove sulfur, before they are burnt in a power station. This means that less sulfur dioxide is formed.

Scientists have also devised a way of removing sulfur dioxide from waste gases (or flue gases) before it can escape from the power-station chimney. This is called **wet scrubbing**.

Sulfur dioxide is an acid gas. Acids react with alkaline chemicals. Wet scrubbing is a process that uses an alkali to react with sulfur dioxide and remove it from flue gases.

Seawater is naturally slightly alkaline and can be used for wet scrubbing. The flue gases are sprayed with seawater droplets, which absorb and react with the sulfur dioxide. The droplets are collected and removed and the cleaned flue gases are released through the power-station chimney.

Other alkalis can be used for wet scrubbing. Powdered lime (calcium oxide) and water can be mixed together to form an alkaline slurry. The flue gases are mixed with air and sprayed with this alkaline slurry. The sulfur dioxide in the flue gases reacts and forms a new solid chemical called calcium sulfate. The solid is collected and removed, and the cleaned gases continue up the chimney. Calcium sulfate can be sold and used as building plaster.

cleaned gases to chimney

waste gases from furnace

calcium sulfate

air

lime and water

Removing sulfur dioxide from flue gases by wet scrubbing with an alkaline slurry.

Particulates

Particulates (tiny particles of carbon and ash) are also found in power-station flue gases. These are a problem because they can make surfaces of buildings dirty, and cause breathing problems. They can be removed by passing them through an electrostatic precipitator. This contains electrically charged plates. The particulates pick up a negative charge, are attracted to the positive plate, and are then collected and removed.

Key word
✓ **wet scrubbing**

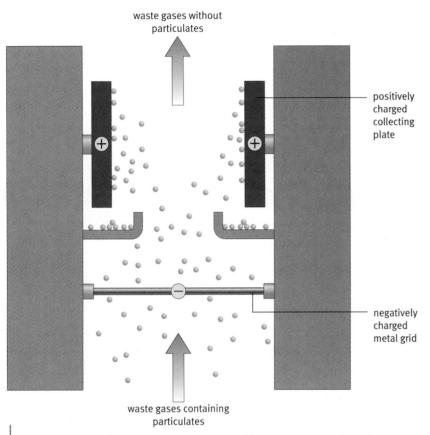

waste gases without particulates

positively charged collecting plate

negatively charged metal grid

waste gases containing particulates

Removing particulates from a power-station chimney to prevent them being released into the atmosphere.

Questions

4   Name three types of fossil fuel that are burned in power stations.

5   What problems are caused by sulfur dioxide in the atmosphere?

6   Describe one way of stopping sulfur dioxide being formed in power stations, and another way of removing it from waste gases if it is formed.

7   Describe how particulates are removed from power-station waste gases.

Science Explanations

Air pollution can affect people's health and the environment. In order to improve air quality it is important to understand where air pollutants come from and how they are made.

- how the Earth's early atmosphere was formed and how it changed as a result of the evolution of photosynthesising organisms
- the importance of the oceans in removing carbon dioxide from the atmosphere leading to the formation of sedimentary rocks and fossil fuels
- which gases now make up the Earth's atmosphere, and how natural events and human activity continue to add gases and particulates to the air
- that burning fossil fuels changes the atmosphere by adding extra carbon dioxide (contributing to global warming) and smaller amounts of other pollutant gases as well as tiny particles of solids (such as particulate carbon)
- that the reactions when fuels burn are oxidation reactions
- that, when a hydrocarbon burns, the oxygen atoms from the air combine with the carbon atoms to form carbon dioxide, and with the hydrogen atoms to form water
- why some pollutants are directly harmful to humans and some are harmful to the environment
- that in any chemical reaction the atoms of the reactants separate and recombine to form different chemical products
- that the number of atoms of each kind is the same in the products as in the reactants
- that the properties of the reactants and products of chemical changes are different
- why the incomplete burning of fuels produces particulate carbon and a poisonous gas, carbon monoxide
- why some fuels produce sulfur dioxide gas when they burn, and why this gas gives rise to acid rain if released into the air
- why the waste gases from fuels burning inside a furnace or engine produce nitrogen oxide gas
- that, when released into the air, nitrogen oxide combines with more oxygen to make nitrogen dioxide gas, which can also contribute to acid rain
- that technological developments such as catalytic converters and wet scrubbing can reduce the amounts of pollutants released into the atmosphere.

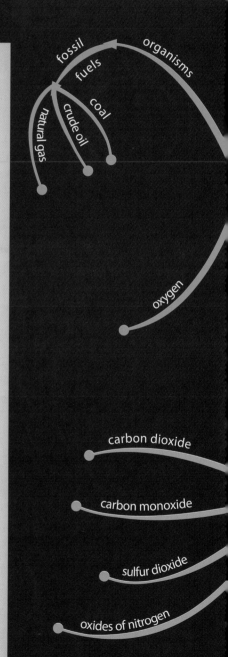

Ideas about Science

Scientists use data, rather than opinions, to justify their explanations. They collect large amounts of data when they investigate the causes and effects of air pollutants. They can never be sure that a measurement tells them the true value of the quantity being measured.

If you take several measurements of the same quantity, the results are likely to vary. This may be because:

- you have to measure several individual examples, for example, exhaust gases from different cars of the same make
- the quantity you are measuring is varying, for example, the level of nitrogen oxides in exhaust gases
- the limitations of the measuring equipment or because of the way you use the equipment.

The best estimate of the true value of a quantity is the mean of several measurements. The true value lies in the spread of values in a set of repeat measurements.

- A measurement may be an outlier if it lies outside the range of the other values in a set of repeat measurements.

- When comparing information on air quality from different places a difference between their means is real if their ranges do not overlap.
- A correlation shows a link between a factor and an outcome, for example, as the level of particulates in the air goes up the number of people suffering from lung disease goes up.
- A correlation does not always mean that the factor causes the outcome.

Scientists publish their results so that their data and claims can be checked by others. Scientific claims are only accepted once they have been evaluated critically by other scientists.

- Reviewers check claims to make sure the scientists who did the work have checked their findings by repeating them.
- The scientific community generally does not accept new claims unless they have been reproduced by other scientists.
- Scientists may come to different conclusions about the same data.

Some applications of science, such as using fuels, can have a negative impact on the quality of life or the environment.

- The only way of producing less carbon dioxide is to burn less fossil fuel.
- Pollution caused by power stations that burn fossil fuels can be reduced by using less electricity, removing sulfur from natural gas and fuel oil, and by removing sulfur dioxide and particulates (carbon and ash) from the flue gases.
- Pollution caused by vehicles can be reduced by burning less fuel in more efficient engines, using low-sulfur fuels, and using catalytic converters.

Review Questions

1 The table shows the percentage of gases in dry air. Copy and complete the table by writing the correct gas next to each percentage. Choose from these gases.

argon   hydrogen
nitrogen   oxygen

Gas	Percentage
	1
	21
	78

2 Ethanol can be used as a fuel instead of petrol. When pure ethanol burns completely, it reacts with oxygen (O_2) and produces carbon dioxide (CO_2) and water (H_2O) as the only products. Copy and complete this drawing to show the products of the reaction when pure ethanol is burned completely.

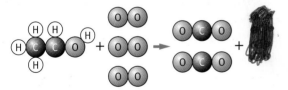

3 Students measured the pH of rain falling on their school playground. They collected and tested six samples of water on the same day. To get a best estimate, they worked out the mean pH from their results.
Their results are shown in the table.
a What is the range?
b What is the mean?

Sample	1	2	3	4	5	6
pH	5.8	5.6	5.8	4.2	5.7	5.6

4 The fumes that are released into the air from a car exhaust cause pollution.
Nitrogen monoxide is made as fuel burns in a car engine. Which chemicals react to make nitrogen monoxide?
- **ammonia**
- **hydrogen**
- **oxygen**
- **carbon dioxide**
- **nitrogen**
- **sulfur**

5 When nitrogen monoxide is released into the air, it reacts with oxygen to form nitrogen dioxide. Why is nitrogen dioxide not formed in the car engine?

6 Which of these measures could do the most to reduce the amount of nitrogen dioxide pollution in the air?
a Use low-sulfur fuels.
b Adjust the balance between public and private transport.
c Encourage people to use less electricity.

7 Several pollutants are released from car exhaust systems. These pollutants do not stay in the air. Over a period of time they are removed by a number of processes. Which pollutants are removed by each of the processes described in the table?

Process	Pollutant
reacts with water and oxygen to form acid rain	
deposited on surfaces, making them dirty	
used by plants and dissolves in sea water	

P1 The Earth in the Universe

Why study the Earth in the Universe?

Many people want to understand more about the Earth and its place in the Universe. The Earth is a very, very small place in a huge and almost empty Universe. How did the substances we are made of come to be here? What is the history of the Universe itself? Natural disasters, such as earthquakes and volcanic eruptions, can be life-threatening. Why do they happen? Can anything be done to predict them?

What you already know

- The Solar System includes planets, asteroids, minor planets, and comets, all orbiting the Sun.

- Some of the planets have moons orbiting them.

- The Sun is a star, one of a vast number of stars in space.

- The speed of a moving object can be calculated using speed = distance/time.

Find out about

- the history of the Universe

- how scientists develop explanations of the Earth and space

- evidence of the Earth's history found in the rocks

- the movement of the Earth's continents

- what seismic waves can tell us about the Earth.

The Science

Science can explain change. The changes that astronomers observe in stars and galaxies can take millions of years. Stars made the atoms found in everything on Earth, including everything in your body. Closer to home, some changes, such as earthquakes, happen very quickly. Others, such as the formation and disappearance of mountain ranges, take millions of years.

Ideas about Science

Scientists depend on data and careful observations of the Earth and Universe. But scientists need to interpret the data they collect. To do this they must use their imaginations. How are scientific ideas tested? There are often arguments between scientists before new data and explanations are accepted.

Find out about

✓ **what is known about the Earth and the Universe**

Our rocky planet was made from the scattered dust of ancient stars. It may or may not be the only place in the whole **Universe** with life.

As the diagrams on these two pages show, scientists know a lot about:
- where and how the **Earth** moves through space
- the history of the Earth.

But there are many things that we still do not know and there are some things we may never know.

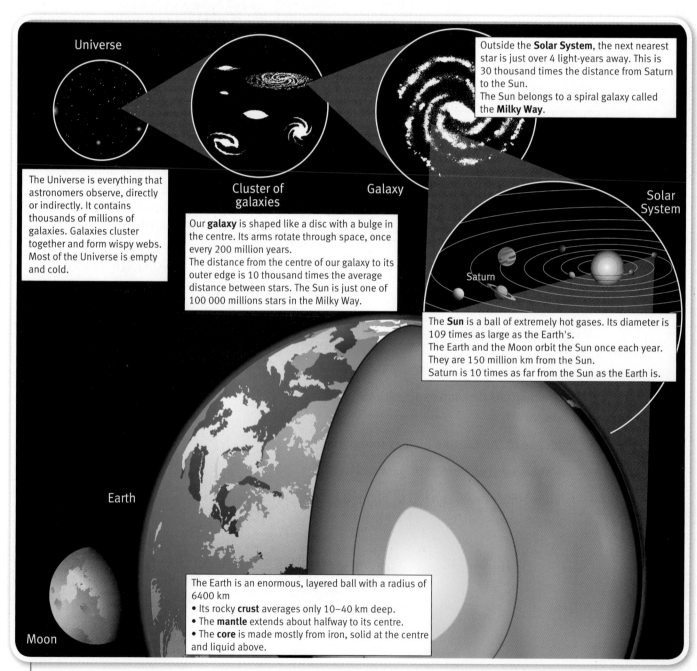

Universe

Outside the **Solar System**, the next nearest star is just over 4 light-years away. This is 30 thousand times the distance from Saturn to the Sun.
The Sun belongs to a spiral galaxy called the **Milky Way**.

The Universe is everything that astronomers observe, directly or indirectly. It contains thousands of millions of galaxies. Galaxies cluster together and form wispy webs. Most of the Universe is empty and cold.

Cluster of galaxies

Galaxy

Solar System

Our **galaxy** is shaped like a disc with a bulge in the centre. Its arms rotate through space, once every 200 million years.
The distance from the centre of our galaxy to its outer edge is 10 thousand times the average distance between stars. The Sun is just one of 100 000 millions stars in the Milky Way.

Saturn

The **Sun** is a ball of extremely hot gases. Its diameter is 109 times as large as the Earth's.
The Earth and the Moon orbit the Sun once each year. They are 150 million km from the Sun.
Saturn is 10 times as far from the Sun as the Earth is.

Earth

Moon

The Earth is an enormous, layered ball with a radius of 6400 km
- Its rocky **crust** averages only 10–40 km deep.
- The **mantle** extends about halfway to its centre.
- The **core** is made mostly from iron, solid at the centre and liquid above.

The Universe is everything there is – from the most distant galaxy to the smallest thing here on Earth.

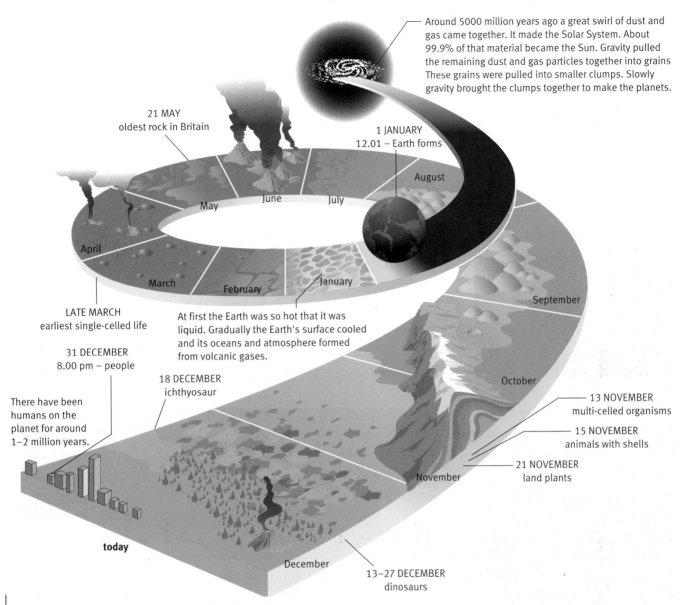

Around 5000 million years ago a great swirl of dust and gas came together. It made the Solar System. About 99.9% of that material became the Sun. Gravity pulled the remaining dust and gas particles together into grains These grains were pulled into smaller clumps. Slowly gravity brought the clumps together to make the planets.

21 MAY
oldest rock in Britain

1 JANUARY
12.01 – Earth forms

August

May
June
July

April

March
February
January

September

LATE MARCH
earliest single-celled life

At first the Earth was so hot that it was liquid. Gradually the Earth's surface cooled and its oceans and atmosphere formed from volcanic gases.

31 DECEMBER
8.00 pm – people

18 DECEMBER
ichthyosaur

October

There have been humans on the planet for around 1–2 million years.

13 NOVEMBER
multi-celled organisms

15 NOVEMBER
animals with shells

21 NOVEMBER
land plants

November

today

December

13–27 DECEMBER
dinosaurs

Timeline: the history of the Earth, scaled as if it took place in one year.

Questions

1 Using the illustration on the previous page as a source, make a list of seven astronomical objects, in order of size. Start with the Moon and end with the Universe.

2 The timeline above shows the age of the Earth.
   a Redraw it as if it happened over a period of 15 years (roughly your lifetime).
   b On this scale, when did life first appear on Earth? When did the dinosaurs die out?

Key words

- ✓ Universe
- ✓ Earth
- ✓ galaxy
- ✓ Solar System
- ✓ Milky Way
- ✓ Sun
- ✓ crust
- ✓ mantle
- ✓ core

Find out about

- ✔ **what makes up the Solar System**
- ✔ **the process that releases energy in the stars and produces new elements**

Astronomers use telescopes of different sorts to observe the night sky. A telescope gathers radiation, such as light, from distant stars. This radiation carries information that astronomers have learned to decode. This helps us to build up our understanding of the Universe and everything in it.

In some places **light pollution**, dust, and dampness in the air stop radiation getting through the atmosphere.

Solar System

The Earth is part of the Solar System. It is one of eight planets that orbit the Sun. Most of these planets have smaller moons in orbit around them. The **dwarf planets** have similar orbits to the planets. Between Mars and Jupiter is the orbit of the small, rocky **asteroids**.

The orbits of the planets and asteroids are roughly circular. **Comets** are large balls of ice and dust. They have very different orbits, as shown in the diagram below.

Astronomers prefer to work in dark places where they will have a clear view of the night sky. They can analyse light from stars to discover how hot they are and what they are made of.

This diagram of the Solar System shows a typical comet orbit. Comets spend a lot of time far from the Sun. They plunge inwards to pass around the Sun before returning to the cold outer reaches of the Solar System.

The Sun

The Sun is the biggest object in the Solar System. It accounts for 99.9% of the mass. It is also the only object that produces its own light. We see everything else by reflected light.

Scientists once struggled to understand the Sun. It could not be a great ball of fire, because fire needs fuel and oxygen and these would have run out long ago.

Then they found that atoms have a central core, called a **nucleus**. Joining small nuclei together releases energy and creates new elements. This process is called nuclear **fusion**.

Nuclear fusion happens only at extremely high temperatures – millions of degrees. This is the temperature of the inside of a star. Hydrogen nuclei fuse to make helium in the Sun.

Heavy elements are made in stars

The most common elements in the Universe are hydrogen and helium. In stars, hydrogen is fused to make helium. Helium is fused to make heavier elements, such as carbon and oxygen.

At the ends of their lives, some big stars explode as supernovae. Their debris, containing all 92 elements, is scattered through space.

The age of the Solar System

As you saw on page 77, the Solar System formed about 5000 million years ago, from a swirling cloud of dust and gas. This process took millions of years.

As it formed, the Solar System gathered debris from a previous generation of dead stars. Except for hydrogen and helium, the chemical elements that make up everything on Earth come from stars. We are made of stardust.

The remains of a star that exploded as a supernova at the end of its life, spreading out into space. One day, this material may be part of new stars and planets.

An artist's impression of the Solar System as it formed, 5 billion years ago. The Sun's gravity pulled in most of the material. A small amount was left in orbit, which formed the planets, their moons, and everything else.

Questions

1 Put these objects in order of size, from the smallest to the biggest:
   a comet   the Moon   the Sun   an asteroid   the Earth

2 Draw a diagram to compare the orbits of the Earth and a comet around the Sun.

3 Why do scientists believe that there must have been stars that existed and died before the Solar System formed?

Key words

- ✓ **light pollution**
- ✓ **asteroid**
- ✓ **comet**
- ✓ **fusion**
- ✓ **dwarf planet**
- ✓ **nucleus**

Find out about

- ✓ how astronomers see into the past
- ✓ ways of measuring the distance to stars

The stars we see in the night sky have a fixed pattern. The Sun, Moon, and planets appear to move against this fixed pattern. This shows that the stars are not part of the Solar System.

Looking back in time

Light moves fast. It could travel the length of Britain in just 6 millionths of a second. At 300 000 km/s, light from the Sun takes just over 8 minutes to reach Earth. This means that you see the Sun as it was 8 minutes ago. You see the stars as they were many years ago.

Although light travels very fast through empty space, it doesn't travel instantaneously from place to place. Its speed is finite.

- Light takes about 1.3 seconds to travel from the Moon to the Earth.
- It takes about 4 years to travel right across the Solar System.
- It takes about 100 000 years to travel right across our galaxy, the Milky Way.

Light-years away

Proxima Centauri is not bright enough to see without a telescope, but it is the closest star outside the Solar System. Light from this star takes 4.22 years to reach Earth. We say that it is 4.22 light-years away.

A **light-year** is a unit of distance used by astronomers. It is the distance travelled by light in one year.

Arcturus is another of the nearer stars. Arcturus is 36.7 light-years away.

How do we know how far away a star is? Astronomers have several methods for measuring the distances to stars. We will look at two of them.

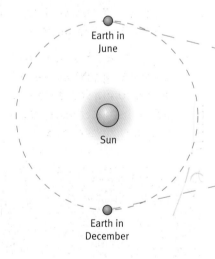

Measuring star distances

Method 1

Parallax: In six months, the Earth moves from one side of the Sun to the other. Seen through a telescope on Earth, a nearby star will shift its position against the background of more distant stars. The nearer a star is, the more it shifts.

This effect is called **parallax**. It provides a way of measuring distance. This method allows astronomers to measure the distance to the nearest stars, including Proxima Centauri.

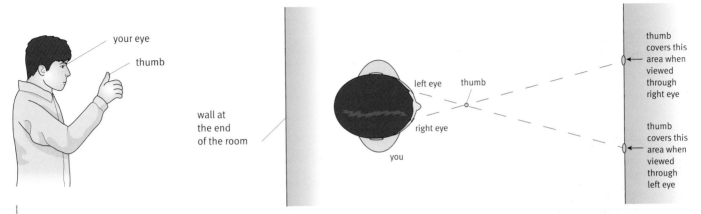

To see the parallax effect, hold up your thumb and look at it with each eye in turn. Your left eye represents the position of the Earth in June, your right eye is the Earth in December.

Method 2

Brightness: The streetlights in the photograph all shine with the same brightness. But the streetlights that are further away appear fainter.

We can use this idea to work out which of two stars is further away. The brighter a star appears to be, the nearer it must be.

But there is a problem with this. We would need to know that the two stars were the same type. Astronomers can analyse the light from the stars to work out how hot they are, and therefore how bright they are. Then we can deduce that, if two stars are the same type, the one that appears fainter must be further away.

Questions

1 It takes light from the Sun eight minutes to reach the Earth.
   a How many seconds is this?
   b Calculate the distance from the Sun to the Earth.

2 Some light from Proxima Centauri is reaching Earth as you read this. How old were you when that light left Proxima Centauri?

3 Suggest why it is easier to use the brightness method of measuring star distances when the observatory is far from any cities or towns.

Key words
- ✔ **light-year**
- ✔ **parallax**

We know that all streetlights shine with equal brightness, so we can deduce that the ones that appear fainter must be further away.

Find out about

- ✔ **our galaxy, the Milky Way**
- ✔ **objects beyond the Milky Way**

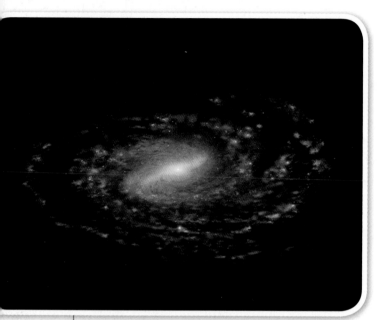

An artist's impression of the Milky Way galaxy, deduced from observations of many millions of stars. The bright yellow dot in the lower arm represents the position of the Sun, although really the Sun is no brighter than other stars.

Our galaxy

Our galaxy is called the Milky Way. A galaxy is a collection of stars. There are thousands of millions of them, held together by gravity. The Milky Way is called this because it looks like a bright band across the night sky. You need to be in a place where there is little light pollution to see it clearly. It is also easier to see the Milky Way from the southern hemisphere.

For over 2000 years, astronomers had suggested that the Milky Way was made up of large numbers of stars. However, this was not confirmed until 1610, when Galileo first used a telescope to look at the night sky.

By counting the numbers of stars in different directions and estimating their distances, astronomers have built up a picture of the galaxy. The illustration on the left gives an idea of its shape.

- The Milky Way has several arms spiralling out from the centre.
- It has a bulge at the centre.
- The Sun is about half-way out from the centre, in one of the spiral arms.

From clouds to galaxies

In the year 964, a Persian astronomer called Abd al-Rahman al-Sufi published a book of his observations. He was the first person to record a faint smudge of light in the constellation of Andromeda. This came to be described as a **nebula**, the Latin word for a cloud.

Over the centuries telescopes improved and many more nebulae were recorded, but no-one could be sure what they were.

- Were they clouds of gas inside the Milky Way?
- Or were they star clusters, far outside the Milky Way?

In 1925, Edwin Hubble, an American astronomer, used a new telescope to try to find out how far away the Andromeda nebula is. The result was surprising. It seemed it was more than a million light-years away, far outside the Milky Way.

Today's space telescopes show that it is indeed a spiral galaxy, similar to the Milky Way. Its name has been changed to the Andromeda Galaxy.

Billions of billions

We now know that there are thousands of millions of galaxies in the Universe, each made up of thousands of millions of stars – perhaps 10^{22} stars in total.

The Hubble Space Telescope image (below) shows some extremely distant galaxies. The light that made this image left its stars over 10 000 million years ago. That was long before the Sun and the rest of the Solar System existed.

Measurement uncertainties

Determining the distances to the stars proved a difficult task for astronomers. First they had to measure the diameter of the Earth's orbit around the Sun. Then they had to use the parallax method to measure the distance to nearby stars. After that astronomers used the brightness method to estimate the distance to further stars. Another method, using variable stars whose brightness varies regularly from day to day, was needed to extend the measurements beyond the Milky Way.

Each of these methods has built-in **assumptions** upon which it depends – for example, the brightness method assumes that two stars that appear similar really are the same type of star.

Each method depended on the results of the previous one, so the final results were, to start with, very **uncertain**. By making many different measurements and comparing the results, astronomers eventually estimated the size of the Universe. And it's very, very big!

A photograph of the Andromeda galaxy. Two other galaxies can be seen in the image: the bright blob above Andromeda and the elliptical smear below it.

An image of distant galaxies, made by light at the end of a long, long journey.

Questions

1 What is the name of the galaxy in which the Solar System is found?

2 What new observation showed that there were objects outside the Milky Way?

3 Roughly how many galaxies are there in the Universe?

4 Explain why measuring the diameter of the Earth's orbit was important for measuring the distance to stars beyond the Solar System.

5 Explain why there are uncertainties about the actual size of galaxies.

Key words
- ✔ nebula
- ✔ assumption
- ✔ uncertain

Find out about

- the age of the Universe
- an explanation called the 'big bang'

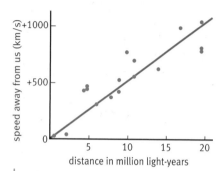

Edwin Hubble published a famous paper in 1929. It showed that more distant galaxies are moving away from us faster.

Imagine yourself on the surface of a very big balloon, looking along a line with galaxies at one-metre intervals. If the balloon is expanding, every metre is growing larger. Let's say the nearest galaxy moves half a metre away from you. In the same time, the second galaxy seems to move away by a metre and the third galaxy by 1.5 metres. The more distant the galaxy, the faster it moves away from you.

The Universe is everything. It is stars and galaxies. It is clouds and oceans. It is bacteria and birds. You are part of the Universe.

A big bang

Until the 20th century, most people thought that the Universe was eternal–it never changed. However, bigger and better telescopes that could see distant galaxies changed all that.

When the astronomers looked at light from distant galaxies, they saw that their spectrum was shifted towards the red end. This is called **redshift**. The amount of redshift shows how fast the galaxies are moving away.

Astronomers had discovered that clusters of galaxies are all moving away from each other. The further they are away, the faster they are moving. The Universe is big and getting bigger. Space itself is expanding.

Scientists now believe that a long time ago the Universe was incredibly hot, tiny, and dense. This explanation is called **big bang** theory.

Testing the theory

The big bang theory passed a major test in 1965. Previously, in 1948, a group of scientists predicted that an afterglow of the big bang event should still fill the whole Universe with microwaves. Years later, two radio engineers in New Jersey noticed an annoying background hiss in their antenna. When they reported this noise, astronomers recognised it as the cosmic microwave background radiation that had been predicted.

The age of the Universe

Picture the expanding Universe – all the galaxies are spreading apart. Now run this imaginary 'film' backwards to see the past history of the Universe – the galaxies all converge at a single point. By working backwards like this, scientists can deduce the age of the Universe.

For their first estimates, 50 years ago, scientists assumed that the galaxies have always moved apart at the same rate as we see today. This may or may not be true. So their answer was only an estimate.

They concluded that the Universe was between 10 000 million and 20 000 million years old. Then, in 2003, new observations of the cosmic microwave background radiation gave a much more precise answer. Scientists estimate that the Universe is 13 700 million years old, plus or minus 200 million years.

Compare this with the age of the Solar System. This was deduced from the age of the oldest rocks on Earth, which are about 4500 million years. The Sun and its planets are about one-third of the age of the Universe.

More evidence for the big bang

Other evidence supports the big bang theory:

- A hot big bang explains why the early Universe was about 76% hydrogen and 24% helium by mass.
- The oldest stars (12 000 million years old) are younger than the Universe. There would be a problem with the theory, if they were older.

Cosmologists at work

Cosmology is the scientific study of the Universe. The big bang is a cosmological theory, devised by cosmologists.

Around the world, there are thousands of cosmologists. Most of them work in universities, usually in groups. When a group develops a new idea, they write a paper for a scientific journal.

Before a scientific paper is published, other experts must first review it. They check it to make sure it has something useful, reliable, and new to say. This process is called **peer review**.

Will the Universe expand for ever?

Cosmologists would like to be able to predict how the Universe will end. The available data is rather uncertain because it is difficult to measure the very large distances to the furthest galaxies, and their speeds.

They would also like to have a better estimate of the mass of the Universe. If the Universe has a high mass, its gravity may cause it eventually to collapse back to a 'big crunch'. However, recent observations of supernovae suggest that the Universe is expanding more and more rapidly. Because the evidence is so uncertain, there are several competing theories of the ultimate fate of the Universe.

Cosmologists make computer models of the Universe. They may have to work with vast amounts of data. This requires a supercomputer, such as the one at Durham University shown here. It can perform 500 billion calculations each second.

Model-independent dark energy test with sigma8 using results from the Wilkinson Microwave Anisotropy Probe

M Kunz, P-S Corasaniti, D Parkinson, and E J Copeland,

Physical Review D **70** 041301 (R) (2004) *ICG 04/30*

Scientific papers are written for other scientists. You have to be an expert to understand them. Through papers like this, cosmologists share their ideas.

Questions

1 List four observations that support the big bang theory.

2 Explain why the discovery of the cosmic background radiation in 1965 was important to the scientists who had proposed the big bang theory.

Key words

- ✓ **big bang**
- ✓ **redshift**
- ✓ **peer review**

Find out about

- ✓ **James Hutton's explanation for the variety of rocks he found**
- ✓ **how old rocks are and how scientists date them**

Around 250 years ago, people started asking new questions about the history of the Earth. They found fossils of seashells and other marine organisms in rocks at the tops of mountains. 'Why here?' they wondered.

James Hutton and the stories that rocks tell

Without some way of building new mountains, erosion would wear the continents flat.

Rivers carry sediment to the oceans, where it settles at the bottom as sand and silt.

Sediments are compressed and cemented to form sedimentary rocks. In some places, layers of sedimentary rocks are tilted or folded.

James Hutton was a well-educated and observant farmer. He watched heavy rains wash valuable soil off farmers' fields. He also noticed that many rocks are made up of eroded material (now called sedimentary rocks). In his mind, he connected these two observations with the idea of a cycle – continents are both eroded and created.

Using the present to interpret the past

In 1785 Hutton explained his startling new theory of the Earth at a meeting of the Royal Society of Edinburgh. At the time this was like a scientific club. The Society published his theory in its *Transactions*, a kind of newsletter. In this way, his ideas reached a much wider European audience.

What Hutton described is the rock cycle. **Erosion** and deposition of sediment take place, very slowly. Over enormous periods of time, these processes add up to huge changes in the Earth's surface. Erosion makes new soil and is therefore essential to human survival. Heating inside the Earth changes rocks and lifts land up.

The Earth has a history. It was not created all at once. The millions of years over which the Earth has changed are called 'deep time'.

Most Europeans in Hutton's time believed that the Earth had been created exactly as they saw it, just 6000 years earlier. This figure for the Earth's age came from an interpretation of the Christian Bible. They rejected Hutton's theory. It took another century and the support of a leading British geologist, Charles Lyell, before Hutton's ideas became accepted.

Dating rocks

Gradually, geologists learned to work out the history recorded in rocks. They used clues like these:

- Deeper is older – in layered rocks, the youngest rocks are usually on top of older ones.
- Fossils are time markers – many species lived at particular times and later became extinct.
- Cross-cutting features – if one type of rock cuts across another rock type, it is younger. For example, hot magma can fill cracks and solidify as rock.

But these clues only tell you which rocks are older than others. They don't tell you how old the rocks are.

Some rocks are radioactive. Scientists today estimate their age by measuring the radiation that these rocks emit. The Earth's oldest rocks were made about 4000 million years ago.

The development of scientific ideas

This first case study about James Hutton, contains examples of:

- data
- explanations
- the role of imagination.

Data

Fossils, rocks of different types, the way that rock types are layered, folded, or joined.

Explanations

Hutton's idea of a rock cycle, different ways of dating rocks.

Imagination

Hutton could imagine the millions of years needed for familiar processes to slowly change the landscape.

Older rocks tend to lie under younger rocks. Different creatures lived at different times in the past; their fossils can help geologists decide when rocks were formed.

Key word

✓ erosion

Questions

1 In what time order did the creatures shown in the picture of the cliff above live? Which layer has the fallen rock come from?

2 Hutton was able to publish his findings in the *Transactions* of the Royal Society of Edinburgh. Why was this important for him?

250 million years ago

Wegener showed how all the continents could once have formed a single continent, called Pangaea.

Key word

- ✔ **continental drift**

Questions

1 In this case study, identify examples of
   a data
   b explanations.

2 'Peer review' involves scientists commenting on the work of other scientists. How did other scientists learn about Wegener's ideas?

3 What were the reasons that other scientists gave for rejecting Wegener's ideas?

How are mountains formed?

A hundred years after Hutton, scientists wanted to know how mountains form. Most geologists believed that the Earth began hot. They compared the Earth with a drying apple, which wrinkles as it shrinks. If the Earth had cooled and shrunk, its surface would have wrinkled too. They claimed that chains of mountains are those wrinkles.

Moving continents?

Scientists discovered radioactivity around 1900. The heating effect of radioactive materials inside the Earth prevents the Earth from cooling. So a new theory of mountain building was needed.

Many people can spot the match between the shapes of South America and Africa. The two continents look like pieces of a jigsaw. Alfred Wegener thought this meant that the continents were moving. They had once been joined together. He looked for evidence, recorded in their rocks.

In 1912 Wegener presented his idea of **continental drift**, and his supporting evidence, to a meeting of the Geological Society of Frankfurt. Geologists around the world read the English translation of his book, *The Origin of Continents and Oceans*, published in 1922.

POLAR EXPLORER DIES

The frozen body of the German meteorologist and polar explorer Alfred Wegener was found on 12 May 1931. Wegener had been leading an expedition in Greenland and went missing just a day after his 50th birthday on 1 November 1930. Unfortunately he is likely to be remembered for being too bold in his science.

Wegener claimed that continents move, by ploughing across the ocean floor. That, he said, explains why there are mountain chains at the edges of continents.

As evidence of continental drift, he found some interesting matches between rocks and fossils on different continents. But most geologists reject such a grand and unlikely explanation for these observations.

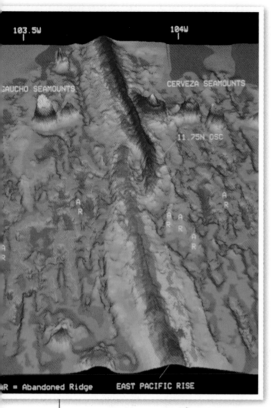

This computer-generated model shows part of the Pacific Ocean floor. (Water is not shown.)

Mapping the seafloor

During the 1950s the US Navy paid for research at three ocean science research centres. The Navy wanted to know how to:

- use magnetism to detect enemy submarines, and
- move its own submarines near the ocean floor, where they could avoid detection.

A few dozen scientists at these three centres, plus two universities, organis ed many expeditions. They gathered huge amounts of data, and published thousands of scientific papers. Their thinking completely changed our understanding of Earth processes.

From stripes in rocks to seafloor spreading

Scientists started to make maps of the ocean floor. To their great surprise, they found a chain of mountains under most oceans. This is now called an **oceanic ridge**. In 1960 a scientist called Harry Hess suggested that the seafloor moves away from either side of an oceanic ridge. This process, called **seafloor spreading**, could move continents.

Beneath a ridge, material from the Earth's solid mantle rises slowly, like warm toffee. As it approaches the ridge, pressure falls. So some of the material melts to form magma. Movements in the mantle, caused by convection, pull the ridge apart, like two conveyor belts. Hot magma erupts and cools to make new rock.

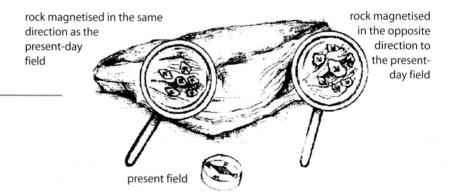

rock magnetised in the same direction as the present-day field

rock magnetised in the opposite direction to the present-day field

present field

Now and again the Earth's magnetic field reverses, for reasons that scientists still do not fully understand. The magnetic north pole becomes the south pole, and vice versa. Iron-rich rocks record the Earth's field at the time that they solidified.

A young British research student, Fred Vine, explained the identical stripe pattern found in rock magnetism on either side of two different oceanic ridges. Vine said that if hot magma rises at a ridge and cools to make new rock, then the rock will be magnetised in the direction of the Earth's field at the time. The science journal Nature published his explanation in 1963.

By 1966 an independent group of scientists had found a clearer pattern of symmetrical stripes in magnetic data either side of a third ridge. This forced other scientists to accept the idea of seafloor spreading.

Tanya Atwater was at university studying geology at that time. She describes a meeting of scientists late in 1966. Fred Vine had shown them an especially clear pattern of magnetic stripes.

'[The pattern] made the case for seafloor spreading. It was as if a bolt of lightning had struck me. My hair stood on end. … Most of the scientists [went into that meeting] believing that continents were fixed, but all came out believing that they move.'

Key words
- ✓ oceanic ridge
- ✓ seafloor spreading

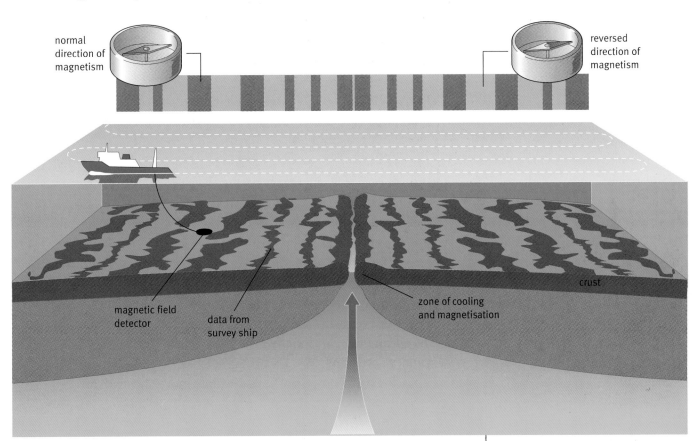

New ocean floor is being made all the time at oceanic ridges. Rock magnetism either side of an oceanic ridge shows the same stripe pattern.

Ocean sediments confirm seafloor spreading

Seafloor drilling in 1969 provided further evidence of seafloor spreading. Sediments further away from oceanic ridges are thicker. This shows that the ocean floor is youngest near oceanic ridges, and oldest far away from ridges.

Questions

4 In this case study, identify examples of:
  a data   b explanations   c predictions

5 Describe carefully how a stripe pattern provides evidence for the seafloor-spreading idea.

Find out about

- ✓ **a big explanation for many Earth processes**
- ✓ **ways to limit the damage caused by volcanoes and earthquakes**

Each red dot on this map represents an earthquake. Earthquakes happen at the boundaries between tectonic plates.

By 1967, seafloor spreading and several other Earth processes were linked together in one big explanation. It was called plate tectonics.

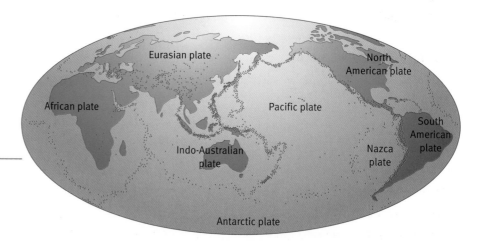

This is the plate tectonics explanation of the Earth's outer layer.

- The Earth's outer layer, or lithosphere, consists of the crust plus the rigid upper mantle.
- It is made up of about a dozen giant slabs of rock, and many smaller ones. These are called **tectonic plates**.
- The lower mantle (below the lithosphere) is hot and soft. It can flow slowly. Currents in the lower mantle carry the plates along.
- The ocean floor continually grows wider at an oceanic ridge by seafloor spreading.
- Ocean floor is destroyed where the plate moves beneath an oceanic trench.
- The result is that the rigid plates move slowly around the surface of the Earth. In places, they move apart. Elsewhere, they push together or slide past one another.

Global Positioning Satellites (GPS) detect the movement of continents. The Atlantic is growing wider by 2.5 cm every year, on average. This is roughly how fast your fingernails grow. In some places, seafloors spread as fast as 20 cm each year.

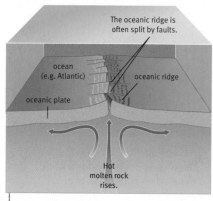

Oceanic ridge.

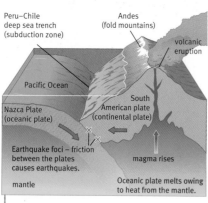

Oceanic trench.

Questions

1 a How far does the Atlantic spread in 100 years (a lifetime)?
   b How far has it spread in 10 000 years (all of human history)?
   c How far has it moved in 100 million years?

2 Today, the Atlantic is about 4500 km wide. How does this compare with your answer to part c?

Plate tectonics explanations

The movement of tectonic plates causes continents to drift. It also explains:

- parts of the **rock cycle**
- mountain-building
- most earthquakes
- most volcanoes

Earthquakes

In some places tectonic plates slide past each other, as at the San Andreas Fault in California.

The shunting of the Earth's plates causes forces to build up along breaks, called fault lines. Eventually the forces are so great that rocks locked together break, and allow plate movement. The ground shakes, making an **earthquake**.

Earthquakes are common at all moving plate boundaries. The most destructive happen at sliding boundaries on land, or undersea, where they may cause a tsunami.

Making mountains

Collisions between tectonic plates cause mountains to be formed. There are three ways that this can happen.

1 Where an ocean plate dives back down into the Earth, volcanic peaks may form at the surface.

2 The pushing movement at destructive margins can also cause rocks to buckle and fold, forming a **mountain chain**.

3 Sometimes an ocean closes completely, and two continents collide in slow motion. The edges of the continents crumple together and pile up, making mountain chains. This is happening today in the Himalayas and Tibet.

> ## Question
>
> 3 Write a step-by-step description of the rock cycle.

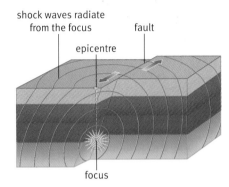

shock waves radiate from the focus
fault
epicentre
focus

Key words
- ✔ **tectonic plates**
- ✔ **rock cycle**
- ✔ **earthquake**
- ✔ **mountain chain**

The rock cycle

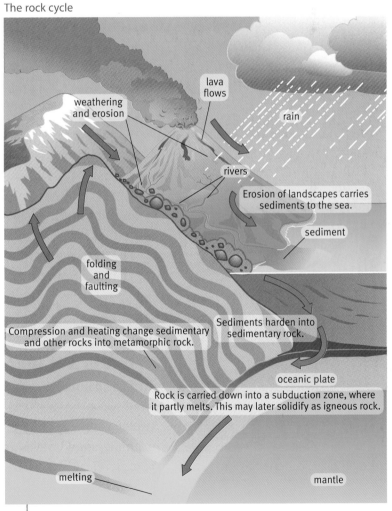

lava flows
weathering and erosion
rain
rivers
Erosion of landscapes carries sediments to the sea.
sediment
folding and faulting
Compression and heating change sedimentary and other rocks into metamorphic rock.
Sediments harden into sedimentary rock.
oceanic plate
Rock is carried down into a subduction zone, where it partly melts. This may later solidify as igneous rock.
melting
mantle

The movement of tectonic plates also plays a part in the rock cycle.

Earthquake damage in Port au Prince, Haiti, following the earthquake of 12 January 2010.

A seismometer – the horizontal and diagonal rods are hinged at the left-hand side, like a garden gate, so that they swing from side to side in the event of an earthquake. This seismometer is designed for use in schools.

Energy of an earthquake

Earth scientists record more than 30 000 earthquakes each year. On average, one of these is hugely destructive.

Vast amounts of energy are released during a powerful earthquake. This energy spreads out from the source of the quake and can be detected at great distances.

Detecting seismic waves

We detect the vibrations of an earthquake using an instrument called a seismometer.

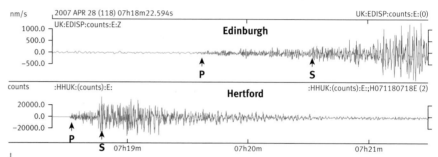

This is a chart, called a seismogram, produced by a seismometer. It shows P-waves and S-waves, which have travelled through the Earth. A seismometer may also detect seismic waves that have travelled around the Earth's surface.

After an earthquake, scientists collect together the data from seismometers in many places around the world to find out as much as possible about the earthquake waves. The picture shows a typical chart, produced when a seismometer detects an earthquake. Two sets of vibrations have been detected:

- The **P-waves** arrive at the detector first (P = primary).
- Then the **S-waves** arrive (S = secondary).

These vibrations that have travelled through the Earth from the site of the earthquake are known as **seismic waves**.

What is a wave?

We are all familiar with waves on the surface of water, but in science the word 'wave' has a special meaning. A **wave** is a repetitive vibration that transfers energy from place to place without transferring matter, for example, sound waves and light waves. In the case of seismic waves, the energy released in an earthquake is carried by the waves around the Earth.

Longitudinal and transverse

P- and S-waves travel differently through the Earth. P-waves travel as a series of compressions (squashed-up regions). In an S-wave, the material of the Earth moves from side to side as the wave travels along.

The pictures show how we can model these two types of wave using a Slinky spring stretched along a table.

To make a P-wave, the end of the spring is pushed back and forth, along the line of the spring. A series of compressions moves along the spring. This type of wave is called a **longitudinal wave**. Sound waves also travel as longitudinal waves.

To make an S-wave, the end of the spring is moved from side to side, at right angles to the line of the spring. This type of wave is called a **transverse wave**. Water waves and waves on a rope are two more examples of transverse waves.

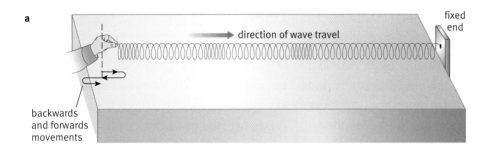

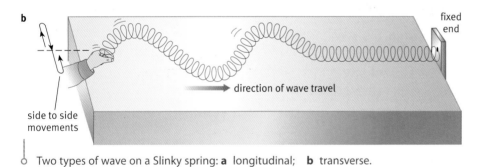

Two types of wave on a Slinky spring: **a** longitudinal; **b** transverse.

Measuring waves

If we draw a simple diagram of a wave, we can define two important quantities:

The **amplitude** is the height of a wave crest above the undisturbed level.

The **wavelength** of the wave is the distance from one crest to the next (or from one trough to the next).

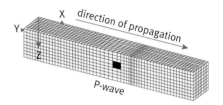

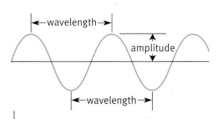

P-waves are sometimes described as 'push-and-pull' waves; S-waves are 'sideways' waves.

Question

1 Look at the seismometer chart on the opposite page. Which waves have a greater amplitude, the P-waves or the S-waves?

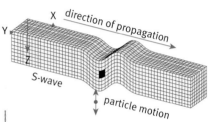

Important information when describing a wave.

Questions

2   Using a ruler, measure the drawings of the Slinky springs (on the previous page) as follows:
    - To find the wavelength of the longitudinal wave, measure the distance from the centre of one compression to the next.
    - To find the wavelength of the transverse wave, measure the distance from the crest of one wave to the next.

3   Which seismic waves have a greater wave speed, P-waves or S-waves? What is the evidence?

4   P-waves travel through the Earth with a wave speed of about 6 km/s. How far will P-waves from an earthquake travel in 5 minutes (300 s)?

Frequency and speed of a wave

Imagine that you are making transverse waves travel along a spring, as in the picture on the previous page. Your hand is the vibrating source of the waves. You can do two things to change the wave.

- Move your hand from side to side by a greater amount. This will give waves with a bigger amplitude.
- Move your hand faster, so that it produces more waves per second. You have increased the frequency of the waves.

The **frequency** of a wave is the number of waves that pass any point each second. It is the same as the number of vibrations per second of the source. Frequency is measured in hertz (Hz). 1 Hz means 1 wave per second.

There is something that you cannot do to the waves on the spring. No matter how you move your hand, you cannot increase their speed. To change the speed of the waves, you would need to use a different spring.

Wave speed is the speed at which each wave crest moves. It is measured in metres per second (m/s).

It is important to realise that frequency and wave speed are two completely different things. The frequency depends on the source – how many times it vibrates every second. Once the wave has left the source, its wave speed depends only on the medium or the material the wave is travelling through.

The distance travelled by a wave can be calculated using the equation:

$$\text{distance} = \text{speed} \times \text{time}$$

Worked example

Measuring distance with a sound wave

The distance-meter emits an ultrasonic sound wave, which bounces from the end of the room and back to the device.

Sound travels at 340 m/s and takes 0.2 seconds to travel the length of the sports hall and back.

How far did the sound travel?

distance = speed × time

distance = 340 m/s × 0.2 s = 68 m

distance = 68 m

How long was the sports hall?

The sound travelled twice the length of the sports hall.

length of sports hall = $\frac{68 \text{ m}}{2}$ = 34 m

The wave equation

Imagine that you are making transverse waves travel along a spring. Moving your hand faster will increase their frequency, as you make more waves per second. The waves look different because their wavelength decreases.

As you know, waves travel out from your hand at a fixed speed. They travel a certain distance in one second. If you make more waves in one second, they will have to be shorter to fit into the same distance.

This shows that there is a link between the frequency of a wave, its wave speed, and its wavelength. Imagine that a source vibrates five times per second. It produces waves with a frequency of 5 Hz. If these have a wavelength of 2 metres in the medium they are travelling through, then every wave moves forward by 10 metres (5 × 2 m) in one second. The wave speed is 10 m/s. In general:

wave speed	=	frequency	×	wavelength
metres per second (m/s)		hertz (Hz)		metres (m)

This link between the three wave quantities applies to all waves of every kind.

The blue waves have a higher frequency than the red waves and so their wavelength is shorter.

Key words

✓ frequency
✓ wave speed

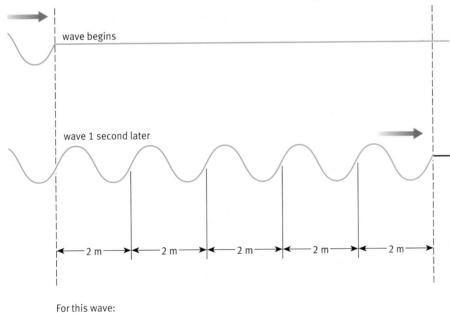

For this wave:

frequency = 5 Hz
So the wave moves on 5 complete wavelengths in 1 second.

wavelength = 2 m
So the wave moves on 10 m in 1 second.

wave speed = frequency × wavelength
= 5 Hz × 2 m
= 10 m/s

The link between wave speed, frequency, and wavelength.

Questions

5 A seismic wave travelling through rock has a frequency of 0.5 Hz. The wavelength is 20 km.
   a Calculate the speed of the wave.
   b The wave passes into a different rock where the speed is 14 km/s. What will the new wavelength be?

6 Look at the red and blue waves at the top of the page. Use the diagram to explain what is meant by 'the wavelength of a wave is inversely proportional to its frequency.'

Find out about

- ✔ **how seismic waves travel through the Earth**
- ✔ **how seismic waves reveal the Earth's internal structure**

A geological scan

Seismic waves are a useful tool for geologists who want to look inside the Earth. They don't have to wait for an earthquake to happen; they can set off small explosions on the Earth's surface and detect the reflected waves. The picture shows how this is used in oil exploration.

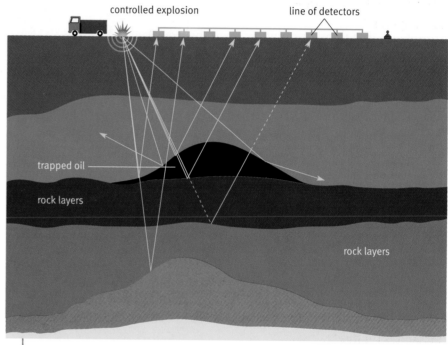

Seismic waves spread out from the explosion and are detected by an array of microphones.

From the pattern of reflected waves, the geologists can work out the structure of the underlying rocks. This is similar to the way in which ultrasound waves are used to produce a scan of an unborn baby.

Inside the Earth

Working on a much larger scale, geologists used seismic waves to discover the inner structure of the Earth. The diagram on the left shows the main layers of the Earth. The thin **crust** is on the outside, then the **mantle**, and finally the **core**. It is impossible to dig down more than a few kilometres into the crust, so how was this picture built up?

In the early years of the twentieth century, scientists had set up seismometers at different sites around the world. This allowed them to detect earthquakes that occurred anywhere on the Earth. They could compare the charts from different sites and work out where and when the quake had happened, as well as how strong it was.

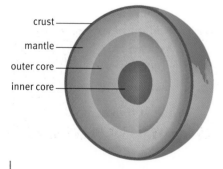

A cross-section through the Earth. This picture is the result of several decades of studies of seismic waves produced by earthquakes.

Shadow zones

Scientists noticed that both P-waves and S-waves reached seismometers close to the earthquake centre, but only P-waves reached seismometers far off, on the other side of the Earth. There was a large 'shadow zone' where S-waves never reached. What could be blocking them?

In 1906, an Irish geologist called Richard Oldham suggested that the Earth had a liquid core. Then, in 1914, a German physicist called Beno Gutenberg published the full answer. He knew that P-waves and S-waves travel differently through the Earth.

- P-waves are longitudinal waves and can travel through solids and liquids (just like sound waves).
- S-waves are transverse waves; they can travel through solids but not through liquids.

Gutenberg realised that it was the liquid core that was blocking the passage of S-waves. From the size of the shadow zone, he was able to work out the size of the core. It is about 7000 km thick, roughly half of the Earth's diameter.

As more sensitive seismometers were set up, it became possible to find out more about the Earth's interior. A Danish scientist, Inge Lehmann, looked at the pattern of P-waves. In 1936, she deduced that there must be a small, solid core inside the liquid core.

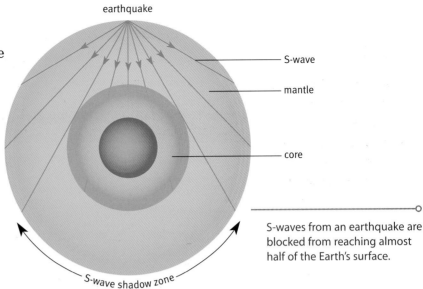

earthquake
S-wave
mantle
core
S-wave shadow zone

S-waves from an earthquake are blocked from reaching almost half of the Earth's surface.

Questions

1 Name the internal layers of the Earth, starting at the centre.

2 Explain how the pattern of S-waves detected at the Earth's surface showed that the Earth must have a liquid core.

3 Suggest how ideas about the structure of the Earth developed when scientists were working in different parts of the world.

Key words

✓ **core**
✓ **mantle**
✓ **crust**

Science Explanations

In this module you will see how scientists gather evidence (data and observations). These data and observations are related to the space beyond Earth – the Solar System, stars, and galaxies – and also to the structure of the Earth and the changes that take place in it.

You should know:

- that the Solar System consists of the Sun, eight planets and their moons, dwarf planets, asteroids, and comets
- that the Sun is one of many millions of stars in the Milky Way galaxy
- how the size of the diameters of the Earth, the Sun, and the Milky Way compare
- that light travels at very high speed and the huge distance it travels in one year (a light-year) is used to measure the enormous distances between stars and between galaxies
- that distant objects in the night sky are observed as younger than they are now because the light now reaching us left them a very long time ago
- that the fusion of hydrogen nuclei is the source of the Sun's energy
- that distant galaxies are moving away from us and that this is because the Universe began as a 'big bang' about 14 000 million years ago
- how the ages of the Universe, the Sun, and the Earth compare
- that the Earth is older than its oldest rocks, which are about 4000 million years old
- that Alfred Wegener's theory of continental drift can explain mountain building
- that seafloor spreading, caused by movements in the mantle, provides evidence for continental drift and tectonic plates
- that the movement of tectonic plates causes earthquakes, volcanoes, and mountain building and contributes to the rock cycle
- that earthquakes produce wave motions on and inside the Earth
- that earthquake waves can be transverse or longitudinal
- that the Earth consists of the inner and outer core, mantle, and crust
- that waves are caused by vibrating sources and the number of waves produced each second is the frequency of the wave
- that waves have amplitude and wavelength, where the amplitude is the distance from the top of a crest (or the bottom of a trough) to the undisturbed position, and the wavelength is the length of one complete cycle
- that wave motion can be described by the equations
  distance = speed × time and wave speed = frequency × wavelength

THE EARTH IN THE UNIVERSE

parallax
brightness
measuring
distances
light-year
stars
expanding Universe
galaxies
Solar System
Sun
planets
moons
comets
asteroids
dwarf planets
our place in the Universe

tectonic plates
light
slow and fast
Earth and Universe
ages
sizes
sense of scale
calculating with data
mean
range
using equations
speed = distance/time
wave properties
wave speed = frequency × wavelength
mathematical skills

evidence in the rocks
core, mantle, and crust
the changing Earth
waves
longitudinal
transverse
frequency
amplitude
wavelength

developing explanation
data
imagination
explanation
predictions
accept
reject
scientific community
disagreement
different possible explanations
different personal experiences
testing new explanations
peer reviewed?
replicated or reproduced?

Ideas about Science

In addition to developing an understanding of the structure of the Earth and the nature of the Solar System, stars, and galaxies, you should understand how scientists develop these ideas and how the work of individual scientists becomes accepted or rejected by the scientific community. The case studies in this chapter illustrate these Ideas about Science.

- New scientific data and explanations become more reliable after other scientists have critically evaluated them. This process is called peer review. Scientists communicate with other scientists through conferences, books, and journals.
- Scientists test new data and explanations by trying to repeat experiments and observations that others have reported.

From these you should be able to identify:

- the statements that are data
- statements that are all or part of an explanation
- data or observations that an explanation can account for
- data or observations that don't agree with an explanation.

Scientific explanations should lead to predictions that can be tested. You should know:

- how observations that agree or disagree with a prediction can make scientists more or less confident about an explanation.

Scientists don't always come to the same conclusion about what some data means. The debate about Wegener's idea of continental drift provides an example of this. You should know:

- why Wegener's explanation was rejected at the time
- that some scientific questions have not been answered yet
- distances to many stars and galaxies are not known exactly because they are so difficult to measure
- the ultimate fate of the Universe is difficult to predict.

Review Questions

1 Besides the Sun, the Solar System contains planets, moons, comets, and asteroids.

Explain the differences between these. You should explain how they move and put them in order of size.

2 Vesto Slipher was an astronomer.

In 1915, he measured the speed of a number of distant galaxies.

These are some of his results.

Galaxy	Speed (km/s)
A	1100
B	500
C	1100
D	600
E	300

Recent work by astronomers shows that these galaxies are moving relative to us with speed given by the graph:

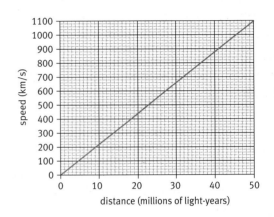

Use the graph to decide which galaxy in Vesto Slipher's table is 23 million light-years away from us.

3 The following statements describe the events leading to an earthquake. They are in the wrong order. Place them in the right order.

a Great pressure builds up where the plates cannot move easily.

b Two tectonic plates meet at a plate boundary at the San Andreas Fault.

c Slow movements of the magma make tectonic plates move.

d Friction at the edges prevents the plates from sliding easily.

e The sudden movement causes an earthquake.

f When the pressure becomes too great, the plates suddenly slip.

4 Copy and complete the following sentences about observing stars.

Choose words from this list.

detection   **galaxies**

light   **planets**

pollution   **sound**

We can only see stars because they give out

....

People in cities find it hard to see stars because of light ....

Astronomers have found that some nearby stars have .... in orbit around them.

B2 Keeping healthy

Why study keeping healthy?

Good health is something everyone wants. Stories about keeping healthy are all around you, for example, news reports about what to eat, how much to drink, new viruses, and 'superbugs'. New evidence is reported every day. So the message about how to stay healthy often changes. It's not always easy to know which advice is best.

What you already know

- Microorganisms can enter the body and cause infections.
- White blood cells defend the body against disease.
- Antibiotics can kill some microorganisms but not viruses.
- Immunisation protects against some diseases.
- Some diseases are caused by unhealthy diet and lack of exercise.
- Scientists work together to investigate and reduce the transmission of infectious disease.

Find out about

- how your body fights infections
- arguments people may have about vaccines
- where 'superbugs' come from
- how new vaccines and drugs are developed and tested
- what causes a heart attack
- how scientists can be sure what causes heart disease
- how your body balances water.

The Science

Some diseases are caused by harmful microorganisms. If you are infected your body has amazing ways of fighting back. Vaccines and drugs can help you survive many diseases, and doctors are always trying to develop new ones. But not all diseases are caused by microorganisms. Your lifestyle may also put you at risk of disease. Media reports often warn about the dangers of smoking, eating badly, and not exercising.

Ideas about Science

So, how do you decide which health reports are reliable? Knowing about correlation and cause and peer review will help. There are also ethical questions (arguments about right and wrong) to consider when deciding how we should use vaccines and drugs.

Find out about

- ✔ **how some microorganisms make you ill**
- ✔ **how bacteria reproduce**
- ✔ **infections**

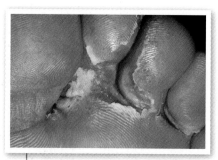

The fungus that causes athlete's foot grows on the skin.

crying, uncomfortable, red gums with white patches on them

cut finger: sore, red cut with pus

swollen glands, runny nose

aching joints, headache, high temperature

monthly check-up

Most days you don't think about your health. It's only when you're ill that you realise how important good health is. Everyone has some health problems during their lives. Usually these are minor – like a cold. But sometimes they can be more serious. Some illnesses may be life-threatening, like heart disease or cancer.

There are lots of reasons for feeling ill. In the doctor's waiting room:

- the man with the painful knee has arthritis
- the young woman feeling sick and tired doesn't know that she's pregnant
- the man having his monthly check-up has had heart disease.

None of these conditions can be passed on to other people. But the other patients all have **infectious** diseases. Infections can be passed from one person to another.

Passing it on

Infections are caused by some **microorganisms** that invade the body. Microorganisms are **viruses**, **bacteria**, and **fungi**.

When disease microorganisms get inside your body, they reproduce very quickly. This causes **symptoms** – the ill feelings you get when you are unwell. Symptoms of infectious diseases can be caused by:

- damage done to your cells when the microorganisms reproduce
- poisons (toxins) made by the microorganisms.

painful, swollen knee joint

sore throat, swollen glands, headache

nausea, tiredness

There are medicines that can cure many diseases caused by bacteria and fungi. But we still don't have many good treatments for diseases caused by viruses. Instead we take medicines that help us feel better until our bodies get rid of the viruses. You will learn more about this later in this chapter.

What are microorganisms like?

Microorganisms are very small. To see bacteria you need a microscope. Viruses are even smaller. They are measured in nanometres, and one nanometre is only one millionth of a millimetre.

Key words

- ✔ **infectious**
- ✔ **microorganisms**
- ✔ **fungi**
- ✔ **viruses**
- ✔ **bacteria**
- ✔ **symptoms**

Microbe attack!

Every breath of air you take has billions of microorganisms in it. And every surface you touch is covered with them. But most of the time you stay fit and healthy. This is because:

- most microorganisms do not cause human diseases
- your body has barriers that keep most microorganisms out.

	Virus	Bacterium	Fungus
Size	20–300 nm	1000–5000 nm	50 000+ nm
Appearance			
Examples of diseases caused	flu, polio, common cold, AIDS, measles	tonsillitis, tuberculosis, plague, cystitis	athlete's foot, thrush, ringworm

Jolene's finger

Jolene cut her finger when she was gardening. She didn't wash it quickly, so bacteria on her skin and in the soil invaded her body. Once inside they started to reproduce. And when bacteria reproduce, they do it in style.

It was just a small cut, so I ignored it. By the time I went to bed it was a bit sore and red. Now it's all swollen and shiny. It really hurts.

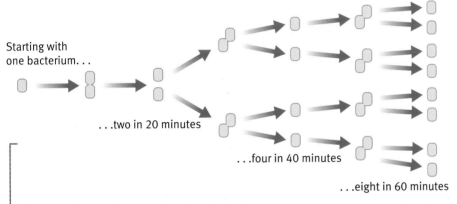

Starting with one bacterium. . .

. . .two in 20 minutes

. . .four in 40 minutes

. . .eight in 60 minutes

Bacteria can reproduce rapidly inside the body.

Reproduction in bacteria is simple. Each bacterium splits into two new ones. These grow for a short time before splitting again. If conditions are right – warmth, nutrients, moisture – they can split every 20 minutes.

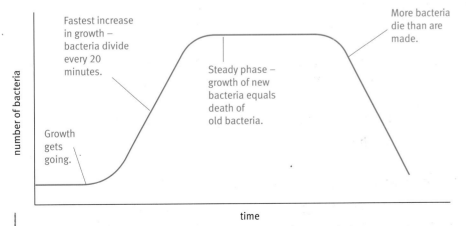

Fastest increase in growth – bacteria divide every 20 minutes.

Growth gets going.

Steady phase – growth of new bacteria equals death of old bacteria.

More bacteria die than are made.

number of bacteria

time

In ideal conditions in a sealed container bacteria can't keep up their fastest growth. Food starts to run out, or waste products kill them off.

Questions

1 Name three types of microorganism that can cause disease.

2 Write down two different diseases caused by each type of microorganism you have named.

3 Explain two ways that microorganisms make you feel ill.

4 What are ideal conditions for bacteria to reproduce?

5 Three harmful bacteria get into a cut. How many might there be after three hours?

Find out about

- ✔ **how white blood cells fight infection**
- ✔ **how you become immune to a disease**

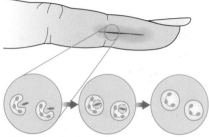

Jolene's body responds by sending more blood to the area.

White blood cells surround the bacteria and digest them.

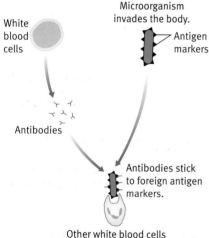

White blood cells

Microorganism invades the body.

Antigen markers

Antibodies

Antibodies stick to foreign antigen markers.

Other white blood cells digest any cell that the antibodies stick to.

One type of white blood cell makes **antibodies** to label microorganisms. A different type digests the microorganisms. All cells have antigen markers on the outside that are unique to that type of cell. The antigen markers on microorganisms are foreign to your body.

The battle for Jolene's finger

Conditions inside Jolene's body are ideal for the bacteria. But they don't have everything their own way.

The redness and swelling in Jolene's finger is called inflammation. Extra blood is being sent to the wounded area, carrying with it the body's main defenders – the **white blood cells**. One type of white blood cell engulfs (surrounds) the bacteria and **digests** them.

The worn-out white blood cells, dead bacteria, and broken cells collect as pus. So redness and pus show that your body is fighting infection. As the bacteria are killed, the inflammation and pus get less until the tissue heals completely.

Your body's army – fighting infection

The parts of your body that fight infections are called your **immune system**. White blood cells are an important part of your immune system.

What's the verdict?

In most cases the body will overcome invading bacteria. Keeping the cut clean and using antiseptic is usually enough treatment. But Jolene's cut is quite deep, so her doctor gives her a course of **antibiotics**. These are antimicrobial chemicals that kill bacteria and fungi but not viruses. Different antibiotics affect different bacteria or fungi.

Everybody needs antibodies – not antibiotics!

A bad cold is something we've all had. And there's not usually much sympathy – 'What's all the fuss about? It's just a cold!'

Natalie has been ill for a few days. Her doctor explains that he won't be giving her any antibiotics. Her cold is caused by a virus, which antibiotics cannot treat. Natalie's own body is fighting the infection by itself.

Fighting the virus

Natalie's neck glands are swollen because millions of new white blood cells are being made there. These white blood cells are fighting the virus in her body.

If antibodies are so good, why do I get ill?

The **antigens** on every microorganism are different. So your body has to make a different antibody for each new kind of microorganism. This takes a few days, so you get ill before your body has destroyed the invaders.

This doesn't really matter for diseases like a cold. But for more serious diseases this is a problem. The disease could kill a person before their body has time to destroy the microorganisms.

Why do you get some diseases just once?

Once your body has made an antibody it can react faster next time. Some of the white blood cells, called **memory cells**, that make the antibody stay in your blood. If the same microorganism invades again, these white blood cells recognise it, reproduce very quickly, and start making the right antibody. This means that the body reacts much faster the second time you meet a particular microorganism. Your body destroys the invaders before they make you feel ill. So you are **immune** to that disease.

Not another one!

Natalie's cold soon got better, but she had only been back at school for about three weeks before she caught another one. If you have an illness like chickenpox, you are very unlikely to catch it again because you are immune. So why do we catch an average of three to five colds every year?

The problem is that there are hundreds of different cold viruses. So every cold you catch is caused by a different virus. To make things worse, the viruses have a very high **mutation** rate. This means that their DNA changes regularly. So do the antigen markers on their surface. The antibody that worked last time will no longer match the marker. Your body needs to make a different antibody to fight the virus. This is why we suffer the symptoms of a cold all over again.

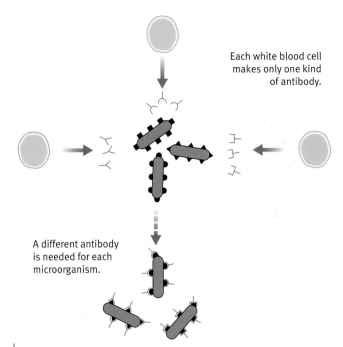

Each white blood cell makes only one kind of antibody.

A different antibody is needed for each microorganism.

Only the correctly shaped antibody can attach to each kind of microrganism.

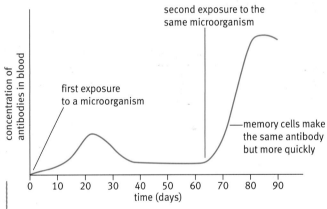

A person is infected twice by a disease microorganism. Their white blood cells make antibodies much faster the second time.

Questions

1 Why are antibiotics not given to patients infected with a virus?

2 Explain two ways that white blood cells protect the body from invading microorganism. You could do this with a diagram.

3 Write down one sentence to describe the job of the immune system.

4 Draw a flowchart to explain how you can become immune to chickenpox.

5 Write a few sentences to explain to Natalie why she will never be immune to catching colds.

Key words

- ✓ **white blood cells**
- ✓ **digests**
- ✓ **immune system**
- ✓ **antibiotics**
- ✓ **antibodies**
- ✓ **antigens**
- ✓ **memory cells**
- ✓ **immune**
- ✓ **mutation**

Find out about

- ✓ **how vaccines work**
- ✓ **deciding if vaccines are safe to use**

In the UK we are lucky to be able to get medicines for many diseases. But it would be even better not to catch a disease in the first place. **Vaccinations** aim to do just that.

Vaccinations make use of the body's own defence system. They kick-start your white blood cells into making antibodies. So you become immune to a disease without having to catch it first.

Small amounts of disease microorganisms are put into your body. Dead or inactive forms are used so you don't get the disease itself. Sometimes just parts of the microorganisms are used.

White blood cells recognise the foreign microorganisms. They make the right antibodies to stick to the microorganisms.

The antibodies make the microorganisms clump together. Other white blood cells digest the clump.

Your body stores some of the white blood cells (memory cells). If you meet the real disease microorganism, the antibodies you need are made very quickly.

The microorganisms are destroyed before they can make you ill. (Not to scale)

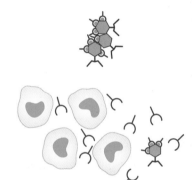

How vaccines work.

Age	Diseases protected against by childhood vaccinations
2, 3 and 4 months	DTB-Hib (diphtheria, tetanus, whooping cough, polio and Hib, a bacterial infection that can cause meningitis and pneumonia) Pneumococcal infection Meningitis C
13 months	MMR (measles, mumps, and rubella)
3–5 years	Diphtheria, tetanus, whooping cough and polio MMR
Girls aged 12–13 years	Cervical cancer caused by human papillomavirus
13–18 years	Diphtheria, tetanus, polio

Many childhood diseases are very rare in the UK because of vaccination programmes.

Are vaccines safe?

Any medical treatment you have should do two things:

- improve your health
- be safe to use.

Vaccines can improve your health by protecting you from disease. They are tested to make sure that they are safe to use. But it is important to remember that no action is ever completely safe. People are genetically different, so they react differently to medical treatments, including vaccines.

Doctors decide that a treatment is safe to use when:

- the risk of serious harmful effects is very small
- the benefits outweigh any risk.

You can read more about this in Section E.

Whose choice is it?

To stop a large outbreak of a disease, almost everyone in the population needs to be vaccinated. If they are not, large numbers of the disease-causing microorganisms will be left in infected people. If the vaccination rate drops just a little, lots of people will get ill.

☐ vaccinated   ■ infected   ■ not vaccinated

The vaccination rate is 98%. Unvaccinated people are unlikely to catch the disease.

The vaccination rate has dropped to 90%. Unvaccinated people are much more likely to catch the disease.

Why does the government encourage vaccinations?

Doctors encourage parents to have their children vaccinated at an early age. In the UK there are mass vaccination programmes for some diseases, such as measles. This means that few people suffer from these diseases. Parents have to balance the possible harm from the disease against the risk of possible side-effects from the vaccine.

- Almost nobody who has a vaccine notices any harmful effects.
- Harmful effects from the MMR vaccine can be mild (3 in every 10 000 children), or produce a serious allergic reaction (1 in every million children).
- Some children who catch measles are left severely disabled (1 in every 4000 cases).
- Measles can be fatal (1 in 10 000 cases).

For society as a whole, vaccination is the best choice. But for each parent it is a difficult choice, with their child at the centre of it. People often perceive the risk of vaccination to be greater than the risk of measles. It is important that people have clear and unbiased information to help them make their decision.

Questions

1 What is a vaccine made of?

2 Describe how a vaccine can stop you from catching an infectious disease.

3 Explain why a vaccine can never be 'completely safe'.

In 1998 the MMR vaccine was wrongly linked to autism. This worried many parents. In the 1970s there were similar worries about the whooping cough vaccine.

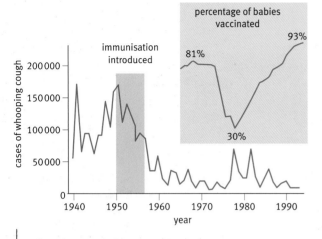

The graph shows the number of whooping cough cases in the UK each year between 1940 and 1992.

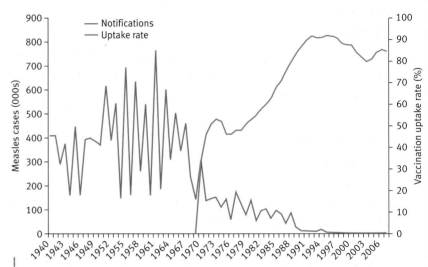

The graph shows how the number of cases of measles changed between 1940 and 2007 and the vaccination uptake since 1970.

Whooping cough:

Disease killed or damaged thousands of children each year.

1950 - vaccine introduced.

1970s - scientific report suggested that there could be a link between the vaccine and serious brain damage in children. Many media stories about the report.

Vaccinations fell from 81% to around 30%.

Over 200000 extra cases of whooping cough in the 1970s and 1980s, with 100 deaths.

1980s - scientists showed that reports of brain damage from the vaccine had been inaccurate. But it took almost 20 years before vaccinations were back to their original level.

Questions

4 To stop a large outbreak of a disease, almost all of the population must be vaccinated against it. Explain why.

5 a Estimate the number of whooping cough cases one year before vaccination began.
   b Describe what happened to the number of cases between 1950 and 1970.
   c What happened to the percentage of babies vaccinated between 1973 and 1979?
   d Explain why this change happened.

6 Look at the number of whooping cough cases between 1965 and 1990. Is there any link with the percentage of babies vaccinated?

7 The number of cases of measles in England and Wales rose to over 5000 in 2008. Suggest a reason for this.

Smallpox

Smallpox was a devastating disease. In the 1950s there were 50 million cases worldwide. This fell to 10–15 million cases by 1967 because of vaccination by some countries. But 60% of the world's population were still at risk.

In 1967 the World Health Organisation (WHO) began a campaign to wipe out smallpox by vaccinating people across the world. In 1977 the last natural case of smallpox was recorded, in Somalia, Eastern Africa.

Should people be forced to have vaccinations?

Governments and public bodies like the National Health Service make decisions about who should be offered vaccinations based on an asessment of risk and benefit. All children are offered the measles vaccine. There is enough measles vaccine for every child in the UK. If everyone had to be vaccinated by law, there would be a much lower risk of any child catching the disease. However, a few children would still get the disease, because vaccinations don't have a 100% success rate.

So it would be possible for measles vaccination to be compulsory – but it isn't. Society does not think it is right to force anyone to have this particular treatment. There is a difference between what *can* be done with science, and what people think *should* be done.

Different decisions

Where you live may make a difference to your choice about vaccination.

People are more likely to catch a disease due to poor hygiene or overcrowded housing. They will also suffer more if they catch a disease because they may:

- be weaker because of poor diet or other diseases
- have less access to medicines and other healthcare.

So people from poorer communities may make different decisions about vaccinations compared with people in better-off communities.

Smallpox killed every fourth victim. It left most survivors with large scars, and many were also blinded.

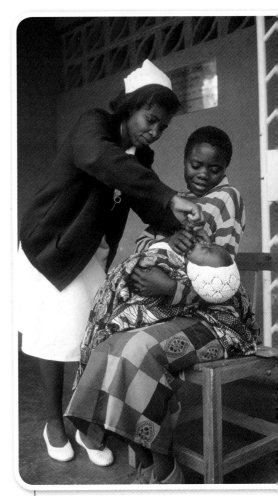

Some people have become concerned about the safety of vaccines for their children. But for many, the decision is easy.

Questions

8 Explain why measles vaccination is not compulsory in the UK.

9 Give two reasons why people in different parts of the world may feel differently about having vaccinations.

10 Scientists cannot make vaccines against every disease. How would you decide which diseases to target?

Find out about

✓ where 'superbugs' come from
✓ how you can help fight them

Microorganisms (bacteria, fungi, and viruses) can be killed by antimicrobial chemicals. Some only inhibit their reproduction. The person's immune system destroys those remaining. Antimicrobials used at home, like bleach, will kill, or inhibit, bacteria, fungi, and viruses. Antifungal chemicals kill, or inhibit, fungi. Antibiotics from the doctor only kill bacteria. They are not effective against viruses.

The first antibiotics

The Ancient Egyptians may have been the first people to use antibiotics. They used to put mouldy bread onto infected wounds. Scientists now know that the mould is a fungus that makes penicillin. In the 1940s scientists started to grow the fungus to make larger amounts of penicillin.

The bugs fight back

To begin with, penicillin was called a 'wonder drug'. Before the 1940s, bacterial infections had killed millions of people every year. Now they could be cured by antibiotics. Antibiotics were also used to treat animals. They were even added to animal feed, to stop farm animals from getting infections.

But within ten years, one type of bacteria was no longer killed by penicillin. It had become resistant. New antibiotics were discovered, but each time resistant bacteria soon developed. The 'superbugs' we are dealing with now are resistant to all known antibiotics, except one. How long before bacteria become resistant to it as well, we don't know.

Where have superbugs come from?

- A tiny change in one gene – a mutation – can turn a bacterial cell into a 'superbug'. Just one superbug on its own won't do much

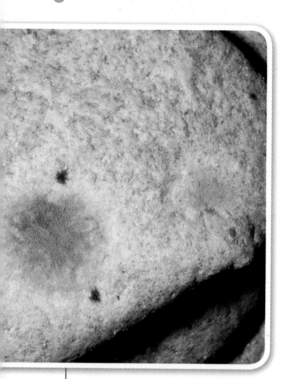

Antibiotics are made naturally by bacteria and fungi to destroy other microorganisms. The fungus growing on this bread makes penicillin.

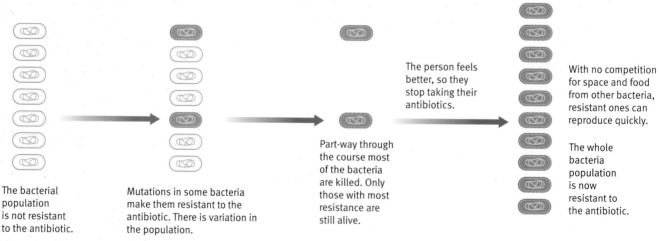

The bacterial population is not resistant to the antibiotic.

Mutations in some bacteria make them resistant to the antibiotic. There is variation in the population.

The person feels better, so they stop taking their antibiotics.

Part-way through the course most of the bacteria are killed. Only those with most resistance are still alive.

With no competition for space and food from other bacteria, resistant ones can reproduce quickly.

The whole bacteria population is now resistant to the antibiotic.

A few mutations can result in antibiotic-resistant bacteria.

damage. But if it reproduces rapidly, it could produce a large population of bacteria, all resistant to an antibiotic. Fungi such as those that cause ringworm and thrush have also become resistant to commonly-used antifungal drugs in exactly the same way.

Why are superbugs developing so quickly?

Two things increase the risk of **antibiotic-resistant** superbugs developing:
- people taking antibiotics they don't really need
- people not finishing their course of antibiotics.

If you are given a course of antibiotics and take them all, it is likely that all the harmful bacteria will be killed. But if you stop taking the antibiotics because you start to feel better, the microorganisms that survive will be those that are most resistant to the antibiotic. They will live to breed another day – and so a population of antibiotic-resistant bacteria soon grows.

How can we stop the superbugs?

Scientists cannot stop antibiotic-resistant bacteria from developing. The mutations that produce these bacteria are part of a natural process. For now, we can only hope that scientists can develop new antibiotics fast enough to keep us one step ahead of the bacteria.

But as well as new drugs, there are other ways of tackling the problem:
- having better hygiene in hospitals to reduce the risk of infection
- only prescribing antibiotics when a person really needs them
- making sure people understand why it is important to finish all their antibiotics (unless side-effects develop).

New drugs in strange places?

Scientists are always looking out for sources of new drugs. For example, crocodile blood might be the source of the next family of antibiotics. A chemical found in crocodile blood is a powerful antibacterial agent. It was discovered by a scientist who wondered why crocodiles didn't die of infections when they bit each other's legs off.

'SUPERBUGS' MRSA ON THE RAMPAGE

These killer bacteria are resistant to almost all known antibiotics. The bad news is that they have broken out of hospitals. People are dying of MRSA 'superbug' infections picked up at work, out shopping, and even at home. And the cause? The very antibiotics we've been using to kill them!

The bacteria MRSA is resistant to almost all antibiotics.

Crocodile blood could be the source of important new antibacterial drugs.

Questions

1 What are antibiotic-resistant bacteria?

2 Write bullet-point notes to explain how antibiotic-resistant bacteria can develop.

3 Describe two things that you can do to reduce the risk of antibiotic-resistant bacteria developing.

Find out about

✓ **how new drugs are developed**
✓ **how they are tested**

Most of us take medicines prescribed by our doctor without asking many questions. We assume that they will do us good. But what if we could ask the scientist who developed the medicine some questions?

Is it safe?

How much should I take?

Are there any side-effects?

How did you discover the drug?

Has it been tested properly?

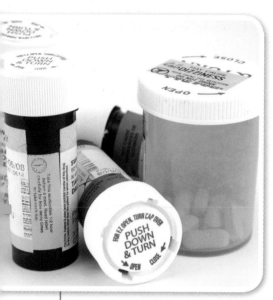

From painkillers to vaccines, antibiotics to antihistamines, medicines are part of everyday life.

Scientists around the world are trying to develop new drugs: new antibiotics, new treatments for asthma and cancer, and new vaccines for malaria and HIV.

Developing a new drug takes years of research, and lots of money. The rewards for a successful discovery can be huge improvements in human health. For drug companies there may also be large profits.

A scientist explains how a new drug is developed:

First we study the disease to understand how it makes people ill. This helps us work out what we need to treat it – for example, a chemical to kill a microorganism, or a chemical to replace one the body isn't making properly.

We search through many natural sources to find a chemical that may be the correct shape to do this. We look at computer models of the molecules to test our ideas.

When we find a chemical that could work, there are many tests that must be done. It's also important that we could make lots of it without too many problems. Only a very small number of possible drugs get through all these stages.

Sian is a cancer research scientist.

Stage 1: human cells

Early tests are done on human cells grown in a laboratory. Scientists try out different concentrations of a possible new drug. They test it on different types of body cells with the disease. These tests check how well the chemical works against the disease – how effective it is. They also give the scientists data about how safe the drug is for the cells.

Stage 2: animal tests

If the drug passes tests on human cells, it is tried on animals. Animal trials are carried out to make sure that the drug works as well in whole animals as it does on cells grown in the laboratory.

Stage 3: clinical trials

If the drug passes animal trials then it can be tested on people. These tests are called **human trials** or **clinical trials**. They give scientists more data about the effectiveness and safety of the drug. Scientists carry out human trials on healthy volunteers to test for safety. They then carry out trials on people with the disease to test for effectiveness and safety. Long-term human trials ensure that the drug is safe and works. It is important that the drug is effective and there are no adverse side effects when used for a long time. These studies may continue after a drug is approved for use, providing more data on the safety of the drug.

Drugs are tested on cells in the laboratory. These are called in vitro tests.

Trials using animals or human volunteers are called in vivo tests.

Not everybody agrees that it is right to test drugs on animals. The British Medical Association (BMA) believes that animal experimentation is necessary at present to develop a better understanding of diseases and how to treat them, but says that alternative methods should be used whenever possible.

If animal trials go well, we apply for a patent. It costs a lot of money to develop a new drug. If we have a patent, no other company can sell the medicine for 20 years. But because clinical trials and the approval process take many years, we often only have about 10 years when we're the only people making the drug.

Key words

- ✓ **human trials**
- ✓ **clinical trials**

Questions

1 Copy and complete the table:

Stage	Testing	To find out
one	Drug is tested on human cells grown in the laboratory.	• how safe the drug is for human cells • how well it works against the disease
two		
three		

2 Developing a new drug is usually very expensive. Suggest why.

Clinical trials – crunch time

Five years ago Anna was diagnosed with breast cancer. Fortunately her treatment worked and she recovered. Now Anna has been asked to take part in the trial of a new drug. Doctors hope it will reduce the risk of the cancer coming back.

What treatment will Anna get?

People who agree to take part in this trial will be put randomly into one of two groups. **Random** assignment into treatment groups is very important in making sure the results of the study are reliable.

One group of people in the trial will be given the new drug, another group will not. This is the **control** group. The results from both groups will be compared.

Anna talks to her doctor:
The problem is I won't know if I'm getting any treatment or not. Could I be risking my health? I know the trial could help people in the future – but what about me? Can you tell me if I will be given the real drug or not?

Before the trial Anna would sign a patient consent form. She signs it to say that all of her questions have been answered. She can also leave the trial at any time. Anyone taking part in a drug trial must give their 'informed consent'.

Anna's doctor wouldn't know if she was getting the new drug or not. Neither would Anna. Someone else would prepare the treatments. This is because Anna would be part of a **double-blind** trial.

If Anna or her doctor knew what treatment she was getting, it could affect the way they report her symptoms. A random double-blind trial is considered the best type of clinical trial.

What treatment will the control group be given?

The drug being tested is a new treatment. In almost all clinical trials the control group are given the treatment that is currently being used. So comparing the results from both groups shows whether the new treatment is an improvement.

Sometimes there isn't any current treatment for an illness. In these cases the control group can be given a **placebo**. This looks exactly like the real treatment but has no drug in it. Using a placebo in a clinical trial is very rare. The control group in Anna's trial will be given a placebo.

Human trials – ethical questions

Taking the placebo would not increase Anna's risk of cancer returning. Taking the new drug may bring other risks. But her doctor will be looking out for any harmful effects. And the new drug may increase her chance of staying well.

It may seem unfair that the control group could miss out on any benefits of the new drug. But remember that not all drugs pass clinical trials. Proper testing is needed to find out if a new drug has real benefits. Tests also give doctors data about the risk of unwanted harmful effects.

- If the trial shows that the risks are too great it will be stopped.
- If the trial shows that the drug has benefits it will immediately be offered to the control group.

Blind trials

In some trials the doctor is told which patients are being given the drug. This may be because they need to look very carefully for certain unwanted harmful effects. The patient still should not know. This method is called a **blind trial**.

Questions

3 Explain why drug trials must be random.

4 Explain the difference between blind and double-blind trials.

5 Describe a situation in which it would be wrong to use placebos in a trial.

6 What do you think Anna should do? Explain why you think this.

double-blind trial

blind trial

open-label trial

Open-label trials

In an **open-label trial** both the patient and the doctor know the treatment. This may be necessary if, for example, a physiotherapy treatment was being compared to a drug treatment, or if a new drug is given to all the patients in a trial. This happens when there is no other treatment and patients are so ill that doctors are sure they will not recover from the illness. The risk of possible harmful effects from the drug is outweighed by the possibility that it could extend their lifespan or be a cure. No one is given a placebo. It would be wrong not to offer the hope of the new drug to all the patients. Penicillin is one example where this happened.

In a drug trial the doctor and/or patient may (✓) or may not (✗) know if the treatment is the new drug.

Key words

- ✓ random
- ✓ control
- ✓ blind trial
- ✓ placebo
- ✓ double-blind trial
- ✓ open-label trial

Find out about

- ✔ **how blood is pumped round your body**
- ✔ **what causes a heart attack**
- ✔ **how to look after your heart**
- ✔ **measuring how hard your heart is working**

Three weeks ago 45-year-old Oliver suffered a serious heart attack. He was very lucky to survive. Now he wants to try and make sure it doesn't happen again.

Your body's supply route

Your heart is a bag of muscle in your body. When you are sitting down it beats at about 70 beats per minute. It has four chambers. The upper two receive blood and the lower two have thick muscular walls to pump the blood. Your heart is a double pump. Tubes carry the blood around your circulatory system.

How blood circulates

Blood enters the right-hand side of your heart from your body. It flows into the right lower chamber, which pumps it to the lungs to pick up oxygen. Your blood then flows back into the upper chamber on the left-hand side of your heart, then into the left lower chamber. There it is pumped to the rest of your body to deliver oxygen. There are valves between the upper and lower chambers to make sure blood flows in the right direction.

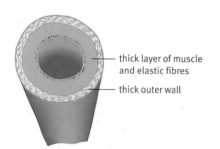

thick layer of muscle and elastic fibres

thick outer wall

Arteries take blood from the heart to your body. The thick outer walls can withstand the high pressure created by the pumping heart.

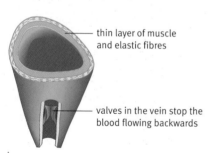

thin layer of muscle and elastic fibres

valves in the vein stop the blood flowing backwards

Veins bring blood back to the heart. The thin layer of muscle and elastic fibres allows the vein to be squashed when you move. This pushes the blood back to the heart.

thin wall (one cell thick) to allow diffusion of oxygen and food to the cells

5–20 μm diameter

Capillaries take blood to and from tissues. The very thin walls (one cell thick) allow oxygen and food to diffuse to cells and waste from cells.

I'll never forget. I went cold and clammy, covered in sweat. And the pain – it wasn't just in my chest. It was down my arm, up my neck and into my jaw. I don't remember much else until I woke up in intensive care. I never want to go through that again.

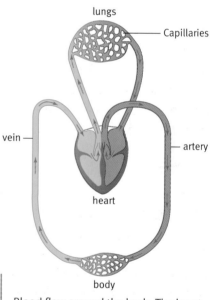

lungs

Capillaries

vein

artery

heart

body

Blood flow around the body. The heart is described as a double pump because the blood passes through the heart twice in each complete circuit.

What is a heart attack?

Blood brings oxygen and food to cells. Cells use these raw materials for a supply of energy. Without energy the heart would stop. So heart muscle cells must have their own blood supply.

Sometimes fat can build up in the coronary arteries. A blood clot can form on the fatty lump. If this blocks an artery, some heart muscle is starved of oxygen. The cells start to die. This is a heart attack.

How serious is the problem of heart disease?

Heart disease is any illness of the heart, for example, a blocked coronary artery and a heart attack.

Oliver survived his heart attack because only a small part of his heart was damaged. He was given treatment to clear the blocked artery. If the blood supply to more of his heart had been blocked, it could have been fatal.

In the UK 230 000 people have a heart attack every year. This is one every two minutes. Coronary heart disease is more common in the UK than in non-industrialised countries. This is because people in the UK do less exercise – most people travel in cars and have machines to do many jobs. And a typical UK diet is high in fat.

What causes heart disease?

Heart attacks are not normally caused by an infection. Your genes, your **lifestyle**, or most likely a mixture of both, all affect whether you suffer a heart attack. There isn't one cause of heart attacks – there are many different **risk factors**. Your own risk of heart disease increases the more of these risk factors you are exposed to.

Is Oliver at risk of another heart attack?

Oliver has a family history of coronary heart disease. He is also overweight, smokes, and often eats high-fat, high-salt food. This diet has given Oliver high blood pressure and high cholesterol levels. All these factors increase his risk of a heart attack. Oliver does like sport – but he'd rather watch it on TV than do exercise himself. Oliver's doctor has given him advice about reducing his risk. This includes advice about drinking alcohol in moderation and reducing his stress levels.

Key words

- ✓ arteries
- ✓ veins
- ✓ capillaries
- ✓ coronary arteries
- ✓ lifestyle
- ✓ risk factors
- ✓ heart disease

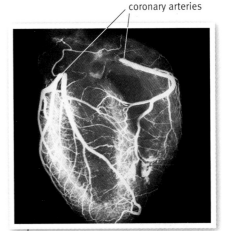

Coronary arteries carry blood to the heart muscle.

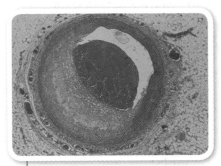

Fat build-up in a coronary artery.

HEALTHY HEART

- Cut down on fatty foods to lower blood cholesterol.
- If you smoke, stop.
- Lose weight to help reduce blood pressure and the strain on your heart.
- Take regular exercise (such as 20 minutes of brisk walking each day) to increase the fitness of the heart.
- Reduce the amount of salt eaten to help lower blood pressure.
- If necessary, take medicines to reduce blood pressure and/or cholesterol level.
- Relax, reduce stress.

Questions

1 Explain why heart cells need a good blood supply.

2 Explain how too much fat in a person's diet can lead to a heart attack.

3 List four lifestyle factors that increase a person's risk of heart disease.

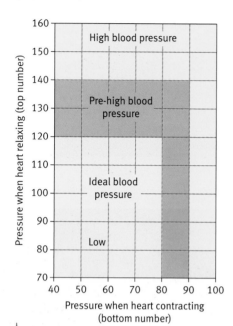

There is a range of 'normal' blood pressure. People differ in height, weight, lifestyle, and sex. All these factors influence what will be normal for that person.

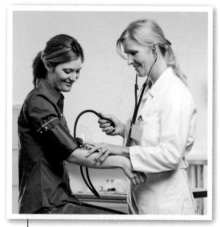

The doctor measures the blood pressure of a patient. High blood pressure can cause heart disease.

Oliver may have been able to prevent his heart attack if he had known his arteries were narrowing. His doctor might have measured his **blood pressure** and **pulse rate**. He could have warned Oliver to live a healthier life.

Monitoring the heart

When arteries become clogged up with fatty deposits, it is more difficult for the blood to be pumped around the body. The heart has to work harder and will have to beat faster.

You can measure how hard your heart is working by measuring your pulse rate. Your pulse is taken on the inside of your wrist or at your neck. You can measure how fast your heart is beating by measuring the beats per minute.

A more accurate measure of how difficult it is for your heart to pump blood around your body is blood pressure. Blood-pressure measurements record the pressure of the blood on the walls of the artery. It is recorded as two numbers, for example, 120/80. The higher number is the pressure when the heart is contracting and the lower number is when the heart is relaxing.

High blood pressure increases the risk of a heart attack. Narrowing of the arteries raises heart rate and blood pressure, and so do drugs such as Ecstasy and cannabis.

Lifestyle diseases

Heart disease and some cancers like lung cancer are **lifestyle diseases**. One hundred years ago infectious diseases killed most people in the UK. Today, better hygiene, vaccinations, and healthcare mean infectious illnesses are more controlled. Lifestyle diseases are much more common than they were. In other parts of the world the lifestyle factors may be different, making other diseases more common.

Questions

4   Heart disease is more common in the UK than in non-industrialised countries. Suggest why.

5   Your neighbour wants to do more exercise, but she gets bored easily and doesn't want to spend money going to the gym. Suggest some ways she could get exercise into her daily life.

It's usually easy for doctors to find the cause of infectious diseases. The microorganism is always in the patient's body. It is harder to find the causes of lifestyle diseases, like heart disease or cancer.

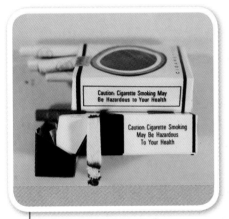

Health warning in 1971.

Health warning in 2003.

Find out about

- ✔ **how scientists identify risk factors for lifestyle diseases**
- ✔ **the evidence needed to prove a causal link**

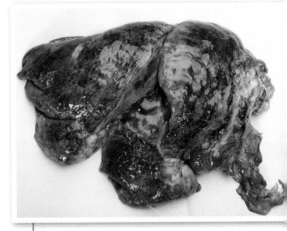

Lung tissue blackened by tar from cigarette smoke.

Smoking and lung cancer

Government health warnings have been printed on cigarette packets since 1971. There was evidence showing a link – a **correlation** – between smoking and lung cancer. But in 2003 the message was made much stronger. How did doctors prove that smoking *caused* lung cancer?

An early clue

In 1948 a medical student in the USA, Ernst Wynder, observed the autopsy of a man who had died of lung cancer. He noticed that the man's lungs were blackened. There was no evidence that the man had been exposed to air pollution from his work. But his wife told Wynder that he had smoked 40 cigarettes a day for 30 years. Wynder knew that one case is not enough to show a link between any two things.

In 1950, two British scientists, Richard Doll and Austin Bradford Hill, started a series of scientific studies. First, they compared people admitted to hospital with lung cancer to another group of people in hospital for other reasons. Smoking was very common at the time, so there were lots of smokers in both groups. But the percentage of smokers in the lung cancer group was much greater.

This data showed a link – a correlation – between smoking and lung cancer. Doll and Hill suggested smoking caused lung cancer. But, a correlation doesn't always mean that one thing causes another.

Cigarettes smoked per day	Number of cases of cancer per 100 000 men
0 – 5	15
6 – 10	40
11 – 15	65
16 – 20	145
21 – 25	160
26 – 30	300
31 – 35	360
36 – 40	415

The data shows how the number of cases of lung cancer in men is affected by the number of cigarettes smoked.

How reliable was the claim?

Doll and Hill published their results in a medical journal so that other scientists could look at them. This is called 'peer review'. Other scientists look at the data and how it was gathered. They look for faults. If they can't find them, then the claim is more reliable.

The claim is also more reliable if other scientists can produce data that suggests the same conclusions.

A major study

In 1951 Doll and Hill started a much larger study. They followed the health of more than 40 000 British doctors for over 50 years. The results were published in 2004 by Doll and another scientist, Richard Peto. They showed that:

- smokers die on average 10 years younger than non-smokers
- stopping smoking at any age reduces this risk

The last piece of the puzzle – an explanation

Lung cancer rates in the USA rose sharply after 1920. The same pattern was seen in the UK.

Many doctors were now convinced that smoking caused lung cancer. But cigarette companies did not agree. They said other factors could have caused the increase in lung cancer, for example, more air pollution from motor vehicles.

The missing piece of the puzzle was an explanation of *how* smoking caused cancer. In 1998 scientists discovered just this. They were able to explain *how* chemicals in cigarette smoke damage cells in the lung, causing cancer. This confirmed that smoking *causes* cancer.

Questions

1. Write down one example of an everyday correlation.

2. Draw a graph to show how the number of cases of lung cancer in men is affected by the number of cigarettes smoked.

3. Explain briefly what happens during 'peer review'.

4. Explain why scientists think it is important that a scientific claim can be repeated by other scientists.

5. It's unlikely that many people would have agreed with Wynder if he'd reported the case he saw in 1948. Suggest two reasons why.

6. If a man smokes 20 cigarettes a day from age 16 to 60, will he definitely develop lung cancer? Explain your answer.

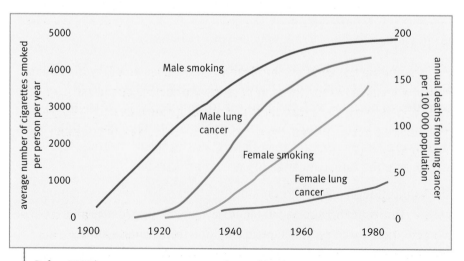

Before 1920 lung cancer was very rare. As smoking became more popular with men, the numbers of lung cancer cases rose. This happened later for women, because very few women smoked until after World War II.

What makes a good study?

There are many reports in the media about studies of health risks. These **epidemiological studies** look for diseases caused by different risk factors. For example, what are the key risk factors for heart disease?

You may want to use this information to make a decision about your own health. So it's important to know if the study has been done well. There are several things you can look for.

How many people were involved in the study?

A good study usually looks at a large sample of people. This means that the results are less likely to be affected by chance.

In the USA a long-term study looked at over 13 000 people across three generations. This study has been hugely important for heart disease research. It has led to the identification of all known major risk factors for heart disease.

How well matched are the people in the study?

Health studies sometimes compare two groups of people. One group has the risk factor, the other doesn't, for example, a study that compares people who exercise with people who do not. In these studies it is important to **match** the people in the two groups as closely as possible.

Genetic studies of heart disease

A large-scale **genetic study** by the Wellcome Trust Case Control Consortium was published in 2007. It showed that heart disease is due to genes and lifestyle.

The research team studied the genomes (types of genes a person has) of 2000 people with coronary heart disease and 3000 healthy controls. They indentified six common alleles or gene variants that are associated with heart disease.

Understanding the genetics that lead to heart disease will help to tell us how much risk a person faces. A person who carries one or more of the 'risk' alleles can still reduce their risk by adopting a healthy lifestyle, monitoring their blood pressure and cholesterol levels, and taking medication.

How big is the risk?

There's one other thing to look for when using data from health studies to make decisions. Imagine a headline like 'Risk of disease is doubled'. It's important to check how big the original risk is. For example, what if the risk of an outcome is that it will happen to one person in a million? An increase of two times is still only two in one million – or one in every 500 000 people. This is still a very small risk.

Looking at the health of lots of people can show scientists the risk factors for different diseases.

Key words
- ✔ **epidemiological studies**
- ✔ **genetic studies**
- ✔ **match**

Questions

7 Name one factor that increases a person's risk of heart disease.

8 Suggest two things you should look for when deciding whether a study was well planned.

9 Your teenage daughter has started smoking. She says 'I don't believe smoking causes heart disease or lung cancer. Grandad has smoked all his life, and he's fine.' How would you explain to her that she may not be so lucky?

Find out about

- ✔ **homeostasis**
- ✔ **why it is important**
- ✔ **negative feedback**

Inside your cells thousands of chemical reactions are happening every second. These reactions are keeping you alive. But for your cells to work properly they need certain conditions. Keeping conditions inside your body the same is called **homeostasis**.

Homeostasis is not easy – lots of things have to happen for your body to 'stay the same'. Look at just a few of the changes happening every second.

Your body works hard to:
- keep the correct levels of water and salt
- control the amounts of nutrients
- get rid of toxic waste products, for example, carbon dioxide and urea.

Control systems

The control systems keeping a steady state in your body work in a similar way to artificial control systems.

All control systems have:
- a **receptor**, which detects the stimuli (the change)
- a **processing centre**, which receives the information and coordinates a response
- an **effector**, which produces an automatic response.

Running has made this athlete hot. His body is sweating more to cool back down. This is an example of homeostasis.

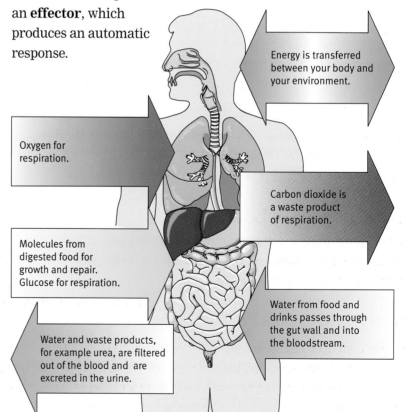

Energy is transferred between your body and your environment.

Oxygen for respiration.

Carbon dioxide is a waste product of respiration.

Molecules from digested food for growth and repair. Glucose for respiration.

Water from food and drinks passes through the gut wall and into the bloodstream.

Water and waste products, for example urea, are filtered out of the blood and are excreted in the urine.

Some of the inputs and outputs that are going on all the time in your body.

How does an incubator work?

Premature babies cannot control their temperature, so they are put in incubators. The incubator is an artificial control system.

An incubator has a temperature sensor, a thermostat with a switch, and a heater. If the temperature in an incubator falls too low, the heater is switched on. The temperature goes up. When the temperature is high enough, the heater is switched off. This type of control is called **negative feedback**:

• any change in the system results in an action that reverses the change.

The diagram below explains how this works.

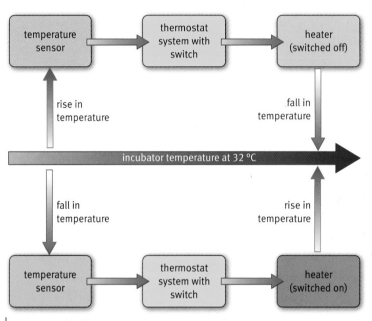

⚬ Negative feedback in an incubator to control temperature

An artificial control system is keeping this baby's temperature steady.

Negative feedback systems are all around you. For example, if the temperature inside your fridge goes up, the motor switches on to cool it down. When it is cool enough, the motor switches off.

What about your body?

Some of the temperature control in your body is automatic too. For example, you do not consciously decide to sweat when you are hot. In the same way, water control is automatic. You do not decide to feel thirsty or to make less urine. These changes are coordinated by both nervous and hormonal communication systems.

Your body also uses negative feedback systems, although your body's systems are more complicated than in the incubator. You have effectors to cool you down and other effectors to warm you up. Negative feedback systems reverse any change to the system's steady state.

Questions

1 Write down a definition for homeostasis.

2 In an incubator, name:
   a a receptor
   b a processing centre
   c an effector.

3 Write down a definition for negative feedback.

Find out about

- ✓ how your body gains and loses water
- ✓ how your kidneys get rid of waste
- ✓ how kidneys balance your water level

Water homeostasis, keeping a steady water level, is done by balancing your body's water inputs and water outputs. The diagram on the opposite page shows how you gain and lose water.

What do your kidneys do?

Your kidneys have two jobs: water homeostasis and **excretion**. Excretion is getting rid of toxic waste products from chemical reactions in your cells. These two jobs are linked because you use water to flush out waste products.

Your **kidneys** control the water balance in your body. They do this by changing the amount of urine that you make. On a hot day, or when you have been running, you lose a lot of water in sweat. So your kidneys make a smaller volume of urine but with the same amount of waste. Your urine will be more concentrated that day.

Getting the water level in cells right is important to maintain the correct chemical concentration levels for cell activity.

More about water balance

Remember that the concentration and volume of your urine varies. On cold days you probably make lots of pale-coloured urine. On hot days you make a smaller volume of darker, more concentrated urine.

The concentration of your blood plasma determines how much water your kidneys reabsorb, and how much you excrete in urine. The concentration of your blood can become higher than normal because of:
- excess sweating because of increased exercise levels
- not drinking enough water
- eating salty food.

In these cases your kidneys will reabsorb more water, making less urine.

Drugs and urine

Some drugs affect the amount of urine a person makes. **Alcohol** causes a greater volume of dilute urine to be produced and can make people very dehydrated. Dehydration can cause dizziness, headaches, and tiredness. Ongoing dehydration can have adverse effects on health including problems with your kidneys, liver, joints, and muscles. Severe dehydration can cause low blood pressure, seizures, increased heart rate, and loss of conciousness.

The drug **Ecstasy** has the opposite effect. It reduces the volume of urine a person makes. It also affects the body's temperature control. Overheating may lead to the person drinking too much water. The amount of water in the body can become dangerously high, causing seizures that can be fatal. Ecstasy also, increases blood pressure and heart rate, increasing the risk of a heart attack.

Experiments carried out with students showed that being able to drink water in the classroom
- increased their concentration time
- improved test results.

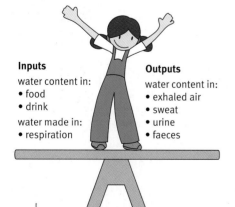

Inputs

water content in:
- food
- drink

water made in:
- respiration

Outputs

water content in:
- exhaled air
- sweat
- urine
- faeces

Water intake and water loss must balance for your body to work well.

Controlling water balance

The control system for water balance is a negative feedback system.

- Receptors in the brain detect any changes in concentration in the blood plasma.
- When the concentration is too high, it triggers the release of a hormone called **ADH** from the **pituitary gland** in the brain. When the concentration is low, no ADH is released.
- The ADH travels in the blood to the kidneys. These are the effectors. ADH affects the amount of water that can be reabsorbed back into the blood. The more ADH, the more water is reabsorbed.

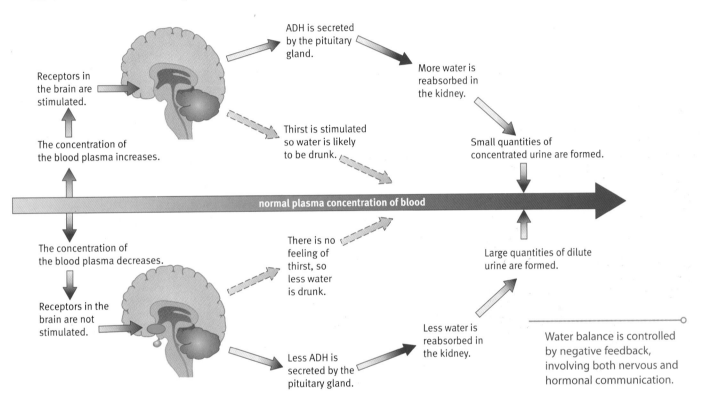

Receptors in the brain are stimulated.

ADH is secreted by the pituitary gland.

More water is reabsorbed in the kidney.

The concentration of the blood plasma increases.

Thirst is stimulated so water is likely to be drunk.

Small quantities of concentrated urine are formed.

normal plasma concentration of blood

The concentration of the blood plasma decreases.

There is no feeling of thirst, so less water is drunk.

Large quantities of dilute urine are formed.

Receptors in the brain are not stimulated.

Less water is reabsorbed in the kidney.

Less ADH is secreted by the pituitary gland.

Water balance is controlled by negative feedback, involving both nervous and hormonal communication.

Drugs affect ADH control

Alcohol and Ecstasy change the volume of urine a person makes because they affect ADH production. Alcohol suppresses ADH production. Less water is reabsorbed in the kidneys, so a larger volume of urine is made. Ecstasy increases ADH production, resulting in a smaller volume of urine as more water is reabsorbed in the kidneys. Body-fluid build-up can lead to brain damage and death.

Questions

1 Copy and complete the table below to show what happens when the concentration of the blood plasma changes.

	Concentration of blood plasma falls	Concentration of blood plasma rises
Pituitary gland secretes	less ADH	
Kidney reabsorbs		
Urine volume		
Urine concentration		increases

2 Ecstasy triggers release of ADH. Explain the effect this will have on water balance.

Science Explanations

Keeping healthy involves maintaining a healthy lifestyle, avoiding infection, using medication when necessary, and our bodies maintaining a constant internal environment.

You should know:

- about microorganisms multiplying rapidly in the human body, damaging cells and releasing toxins to produce disease symptoms
- how white blood cells engulf and digest microorganisms or produce antibodies to destroy them
- how specific antibodies recognise different microorganisms
- how memory cells provide the body with immunity by making specific antibodies rapidly if the body is re-infected, destroying the microorganisms
- how a safe form of a microorganism, called a vaccine, causes the body to produce antibodies
- that drugs and vaccines can never be completely safe because people are different genetically and react differently
- that antimicrobials are used to kill bacteria, fungi, and viruses
- how mutations in microorganisms can make them resistant to antimicrobials
- how new drugs are tested on animals, human cells, and healthy volunteers, then ill people for safety and effectiveness
- about long-term drug trials, including open-label, blind, and double-blind trials
- why using placebos in human trials raises ethical issues
- that the circulatory system consists of the heart, which is a double pump with its own blood supply, arteries, veins, and capillaries
- how fatty deposits in blood vessels can trigger heart disease
- how genetic and lifestyle factors, such as diet, exercise, stress, smoking, and misuse of drugs, can cause heart disease
- how to measure a person's pulse rate
- that high blood pressure may indicate heart disease
- how nervous and hormonal systems help maintain homeostasis
- how body systems detect stimuli with receptors, coordinate responses with a processing centre, and produce a response with effectors
- how negative feedback between receptors and effectors helps to maintain homeostasis
- how kidneys regulate water balance by producing dilute or concentrated urine
- how the hormone ADH controls urine concentration through negative feedback and how drugs affect ADH.

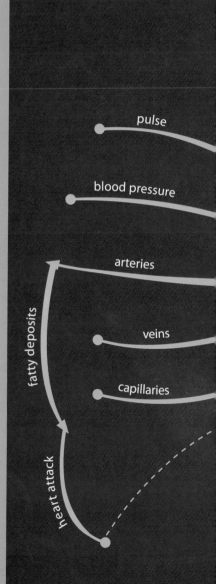

pulse

blood pressure

arteries

fatty deposits

veins

capillaries

heart attack

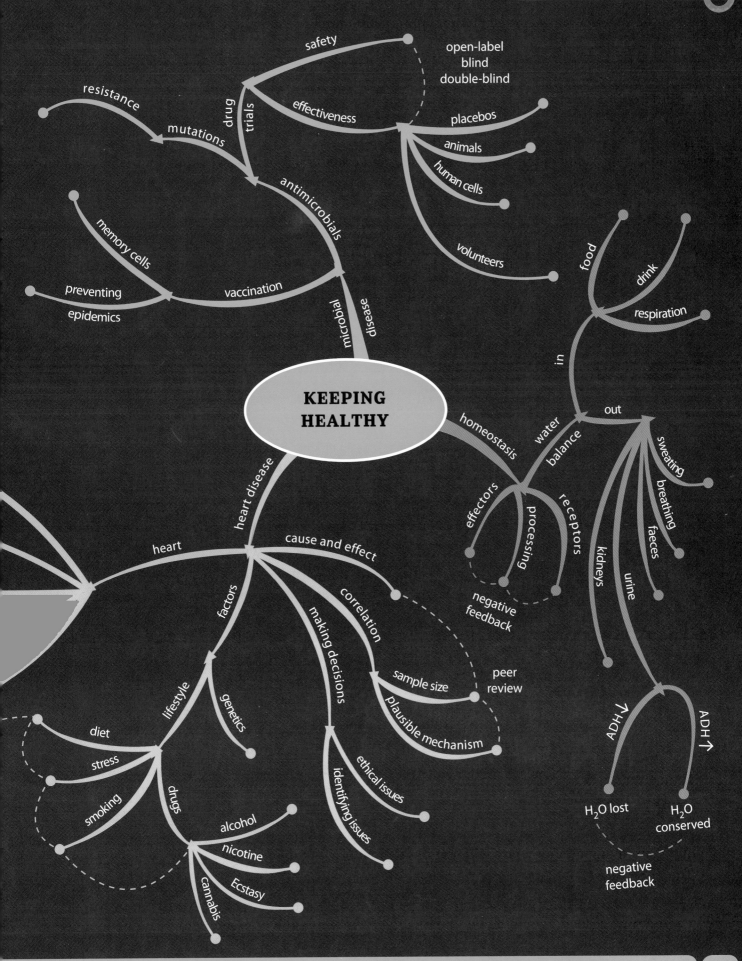

KEEPING HEALTHY

safety

open-label
blind
double-blind

resistance

drug
trials

effectiveness

placebos

mutations

animals

antimicrobials

human cells

memory cells

volunteers

preventing

food

epidemics

vaccination

drink

respiration

microbial

disease

in

water

homeostasis

out

balance

sweating

heart disease

effectors

processing

receptors

breathing

heart

cause and effect

negative
feedback

faeces

factors

correlation

kidneys

urine

making decisions

sample size

peer
review

lifestyle

genetics

plausible mechanism

ADH↓

ADH↑

diet

stress

ethical issues

smoking

drugs

identifying issues

H₂O lost

H₂O
conserved

alcohol

nicotine

negative
feedback

cannabis

Ecstasy

Ideas about Science

This module provides opportunities to develop your understanding of cause–effect explanations, how scientists share their ideas, and how decisions about scientific issues are made, including ideas about risk.

If an outcome increases or decreases as an input variable increases there is a correlation between the two.

You should be able to:

- suggest and explain an example of a correlation from everyday life, such as an increase in the number of cigarettes smoked increases the risk of developing heart disease
- identify a correlation when given data such as text, a graph, or a table
- understand that a correlation does not prove a cause and that the outcome might be caused by some other factor, for example, icecream sales increase as hayfever increases, but icecream does not cause hayfever.

Scientists investigate claims that a factor increases the probability of an outcome, such as the link between cigarette smoking and heart disease, by closely matching different groups of the population or choosing them randomly. You should be able to critically evaluate such studies by commenting on sample size and how well the samples are selected or matched.

Even when evidence exists that a factor is correlated to an outcome, scientists look for a causal mechanism. For example, smoking increases the effect of heart disease because of the effects of nicotine on the body. Nicotine is the mechanism.

Scientists report all their claims to scientific conferences or scientific journals. This is so other scientists can peer review the evidence and claims.

This gives the claim credibility, especially when the findings have been replicated by another scientist.

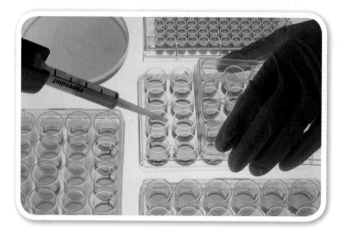

Some questions cannot be answered by science, for example, those involving values. You will need to be able to distinguish questions that can be answered by using a scientific approach from those that cannot, such as should vaccinations be compulsory?

When discussing these questions, the benefits and the size of the perceived and measured risk must be considered.

Some forms of scientific work have ethical implications that some people will agree with and others will not. When an ethical issue is involved, you need to be able to:

- state clearly what the issue is
- summarise the different views that people might hold.

When discussing ethical issues, common arguments are that:

- the right decision is the one that leads to the best outcome for the majority of the people involved
- certain actions are right or wrong whatever the consequences, and wrong actions can never be justified.

You will need to be able to:

- identify examples based on both of these statements.

Review Questions

1 Our bodies are sometimes invaded by microorganisms. We can protect ourselves by having a vaccination containing dead microorganisms. Some of the statements below describe how vaccination helps to protect us from disease-causing microorganisms.
Use only the correct statements and place them in the correct order.

a The disease-causing microorganisms are destroyed.

b White blood cells produce lots of antigens.

c Memory cells rapidly make antibodies for the disease.

d The body slowly makes antibodies to the disease.

e The disease-causing microorganisms enter the body.

f The disease releases antibodies into the blood.

g We receive a vaccination against the disease.

2 New drugs are tested for effectiveness and safety. Explain what and who the drugs are tested on and the role played by open-label, blind, and double-blind trials.

3 Doctors have to make decisions about when to use placebos and who they should be used on. Explain what placebos are, why they are used, and when they should not be used.

4 Kidneys regulate the amount of water present in the human body.

a Explain the role played by ADH in the functioning of the kidneys and how ADH secretion is controlled by negative feedback.

b Explain the effects of:

i alcohol on the functioning of the kidney

ii Ecstasy on the functioning of the kidney.

5 Eating a diet containing a lot of fatty food can increase the risk of heart disease.
Different people have different views about this.

To answer these questions, you may use each person once, more than once, or not at all.

Jane
I read that eating fatty foods will cause heart disease. But I believe scientists who say it will just increase my risk of developing heart disease.

Ranjit
My grandad ate fatty food all his life. He died of influenza at 83. Scientists look at lots of data before they conclude that a high-fat diet increases the risk of heart disease.

Peter
We only know that fatty foods can cause heart disease because many scientists have collected data. If there was only one study they would be less sure.

Stella
I am a food scientist. My findings are always checked by other scientists before they are published.

a Which person says that the absence of replication is a reason for questioning a scientific claim?

b Which person is suggesting that individual cases do not provide convincing evidence for or against a correlation?

c Which person is describing the process of peer review?

d Which **two** people are suggesting that factors might increase the chance of an outcome but not always lead to it?

C2 Material choices

Why study material choices?

All the things we buy are made of 'stuff'. That stuff must come from somewhere. Before about 1900, virtually everything we used was made of naturally occurring materials that came from plants, animals, or rocks. Since then, people have discovered ways to change the properties of some materials that occur in nature – and also to make completely new materials. Materials are chosen for specific jobs because of their properties.

What you already know

- Chemicals can be elements or compounds.
- Chemicals can also be mixtures – two or more chemicals mixed together but not chemically combined.
- Different chemicals and materials have different properties.
- The properties of a material make it suitable for particular uses.
- Molecules are made of atoms.
- In a chemical reaction, atoms separate and recombine to form different chemicals.
- In a chemical reaction, the atoms are conserved.
- How to analyse the results of an experiment.

Find out about

- the testing and measurement that helps people to make good choices when buying products
- some of the explanations scientists use to design better materials
- the variety of polymers and plastics, and how they are used to meet our needs
- how nanotechnology is helping scientists to design new materials with a wide range of properties.

The Science

Scientists use their knowledge of molecules to explain why different materials behave in different ways. This gives them the ability to design new materials to meet a wide range of needs.

Ideas about Science

Scientists test products to check that they can do the job, are good value, and safe. You can use data from these tests when you buy a product. So you need to be able to judge whether or not the results can be trusted.

135

Find out about

- ✓ materials and their properties
- ✓ natural and synthetic materials
- ✓ long-chain polymers

Key words

- ✓ material
- ✓ properties
- ✓ polymer
- ✓ natural
- ✓ synthetic
- ✓ ceramic
- ✓ metal
- ✓ mixture
- ✓ flexible

The newest fashion CHOCOLATE SHOES

Glossy shoes with a perfect fit. No rubbing, no corns, making your feet comfy, tasty, and attractive. Just dip you feet in luxurious molten chocolate to form a close-fitting shoe in one of three gorgeous colours: milk, dark, or white.

Latex is a natural polymer that can be tapped from rubber trees. After treatment, it is used in a wide variety of products, including the soles of shoes.

What the advertising agency didn't tell you

Of course chocolate shoes are a joke. Chocolate is not a good **material** for making shoes. Here are some reasons:

- chocolate would crack
- it would melt in warm weather
- dogs would follow you and lick your feet
- it would wear away too quickly
- it would leave a mess on the carpet.

Maybe not chocolate

Although chocolate does not have the right **properties**, the idea of moulded shoes is not new. South American Indians used to dip their feet in liquid latex straight from the rubber tree. They would sit in the sun to let the latex harden, forming the first, snugly fitting, wellies. Latex is more suitable than chocolate for making shoes. Let's see what properties it has that make it better.

Fantastic elastic

The most obvious difference between latex and chocolate is that latex is **flexible**. Any material chosen to make our shoes needs to be flexible so you can bend your feet. It also needs to be:

* hard wearing because you will walk on it
* waterproof
* a solid at room temperature
* elastic so it keeps its shape
* tough so that it won't crack when it bends.

Latex has all these properties whereas chocolate does not. Latex is a polymer.

What are polymers?

All **polymers** have one thing in common. Their molecules are very long chains of atoms. This is true for **natural** polymers such as cotton, leather, and wool and for **synthetic** polymers such as polythene, nylon, and neoprene. Most of the materials used to make shoes are polymers.

Choosing materials

We make items from a huge range of materials. We use **ceramics** for mugs, plates, tiles, glass windows, bricks, and toilets. We use **metals** for an enormous number of products including aircraft, cars, pipes, wires, jewellery and sports equipment. We use polymers for making bags, clothes, window frames, and computers. All these materials are chemicals. Some of the metals are pure chemicals, not mixed with anything else; most of the materials we use are **mixtures** of chemicals.

When a designer is deciding how to make a product they can choose which material to use according to which properties they require the finished item to have.

What's in a name?

Sometimes words can have more than one meaning. The word 'material' can mean cloth or fabric, but to a scientist it means any sort of stuff you can use to make things from.

Products are made from a wide range of materials.

Questions

1 Look at the picture of young people in a car. Identify items that could be made from:
   a ceramics
   b metals
   c polymers.

2 Leather is a natural polymer. Suggest which properties of leather make it a good material for smart shoes.

3 Steel is a metal. Steel is sometimes used to make toecaps for work boots, but it is not used to make the whole boot. Suggest some properties that steel does *not* have that would be useful in a boot.

Find out about

- ✔ how natural and synthetic polymers meet our needs
- ✔ examples of polymers and their uses
- ✔ how and why natural polymers are being replaced with synthetic ones

All sorts of polymers meet people's most basic needs. These include:

- physical needs for shelter, warmth, and transport
- bodily needs for food, water, hygiene, and healthcare
- social and emotional needs for human contact, leisure, and entertainment
- needs of the mind to stimulate thinking and creativity.

Natural polymers

Before synthetic polymers were discovered, these needs were met using natural polymers as well as materials like metals, glass, and ceramics. The pictures below show some materials that are natural polymers.

Cotton fabric has been used for clothing and household textiles for many years. Cotton grows around the seeds of the cotton plant.

Silk looks and feels luxurious, and has long been popular for high-quality clothing. It is a protein fibre obtained from the cocoons of the silkworm.

Wool is good at regulating temperature, so is used in a variety of clothing and textile products. It is a protein fibre obtained from animals such as sheep.

Doctors and other health workers wear gloves made of natural **rubber** (latex) for protection and to prevent infection. Latex is tapped from some plants, including the rubber tree.

Fur was one of the first materials used for clothing. Today its use is controversial – many view its use as cruel and unnecessary.

Paper has many uses including for decoration, for leisure, and for passing on information. It is made from wood pulp, which consists of cellulose fibres.

From natural to synthetic

The pictures below and on the following page show some synthetic polymers. Many items that used to be made from natural polymers are now made of synthetic ones. As more and more synthetic polymers are made, designers and engineers have more to choose from when they make new products. Each new polymer has different properties, which may be superior to those available in nature.

- Natural fibres for clothing are replaced with synthetic fibres, which may be easier to wash, hold their shape better, or be available in a wider range of colours.
- Fur and leather may be replaced with synthetic polymers, which avoids using animal products.
- Wood is often replaced with plastics, which are much lighter and do not rot or require painting.
- Paper bags are often replaced with plastic ones, which are lighter and also waterproof.

Polythene bags help people to protect, store, and carry food.

This patient in Sri Lanka is fitting a new leg made of polypropylene.

The world's first inflatable church made from PVC.

Polyester is used to make hulls and sails.

PET is a polyester used to make soft-drinks bottles and other food containers.

This acrylic painting was on show in a shop in Zanzibar.

Manchester City stadium roof is made from polycarbonate 'glass'.

Kevlar helmets have saved many soldiers' lives.

A wet suit made from neoprene offers warmth and protection.

Questions

1 Create a chart, diagram, or table to show how polymers can meet our needs. Use the examples throughout Section B and any other examples that you know of.

2 Give at least three examples of objects that were once made of natural polymers but are now made of synthetic ones. In each case, suggest a reason for the change.

3 Name an object now made from plastic rather than metal and explain why.

4 Name an object now made from synthetic polymers rather than ceramics and explain why.

5 Give an example where replacing natural polymers with synthetic ones may be good for:
   a the environment
   b animal welfare.

6 Give an example where replacing natural polymers with synthetic polymers may be bad for the environment.

Getting the right material

Manufacturers and designers have to choose the right materials to make their products. They decide which materials to use based on their properties and cost. In many products, the materials include polymers.

Find out about

- ✔ **words scientists use to describe materials**
- ✔ **testing materials to ensure quality and safety**

Key words

- ✔ **strong**
- ✔ **tension**
- ✔ **compression**

Modern materials with special properties.

For example, the soles of shoes have to be flexible, hard wearing, and strong. Also, they must not crack when they bend – they have to be tough. A synthetic rubber is a good choice.

The case of a computer has to be very different from shoe soles. It needs to be stiff, strong, and tough. People want a case that resists scratches and keeps its appearance. So the polymer has to be hard.

Material words

When scientists describe the properties of materials, they use special words. Some of these, like 'strong', have everyday meanings that are similar to their technical meaning. However, some are a little different.

A material is **strong** if it takes a large force to break it. Some materials are strong when stretched. Examples are steel and nylon, which are strong in **tension**. Concrete tends to crack when in tension but it is very strong in **compression**. This makes it useful for pillars and foundations.

A suspension bridge must be strong. The cables are made of steel, a material that is strong in tension. The columns are made of concrete, a material that is strong in compression.

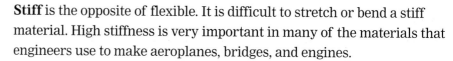

Key words
- ✓ stiff
- ✓ hard
- ✓ soft
- ✓ density

A testing machine for plastic packaging. Measuring the force needed to crush the container gives a value for the strength of the pack.

Stiff is the opposite of flexible. It is difficult to stretch or bend a stiff material. High stiffness is very important in many of the materials that engineers use to make aeroplanes, bridges, and engines.

Hard and **soft** are also opposites. The softer a material, the easier it is to scratch it. A harder material will always scratch a softer one.

In many applications it is also important to know how heavy a material is for its volume. Materials such as steel and concrete have a high **density**. Other materials are very light for their volume and have a low density. Examples are foam rubber and expanded polystyrene.

Measuring the words

Technical words help to describe materials. There are times when more than a description is needed. Accurate measurements of properties are necessary when it is important to compare materials and test their quality.

For example, a pole used for pole vaulting must be flexible, but not too flexible or it will not support the weight of the person using it. The landing pit must be soft but not too soft so that the pole vaulter lands safely. In situations like these the properties of the materials must be measured to make sure that they are just right.

Engineers test materials and products and measure their properties. The flexibility of a material can be found by measuring how much it bends under a given force. The force that breaks a material tells you its strength. The density of a material can be calculated from its mass and volume.

This machine measures the force needed to break sewing threads. Samples from every batch that leaves the factory are tested to make sure the threads are always the same.

Questions

1 Look at the picture of rollerbladers. Identify items that they are wearing, or parts of the rollerblades, that are:
   a flexible   b stiff   c strong   d hard

2 Look at the two pictures of material testing. Which is a test of strength in tension? Which is a test of strength in compression?

3 Suggest reasons for measuring the strength of packaging materials.

Quality control

It is particularly important to take accurate measurements when someone's safety depends on it.

Abseiling ropes must not break when they are being used, so the ropes are tested to make sure that they are strong and safe. The ropes must be able to hold much more than the weight of one person.

Most ropes are produced in batches. A machine is used to test a sample of rope from each batch and find out the force that is needed to break it. Standard procedures are followed carefully to make sure the measurements are accurate. **Accuracy** is how close a measurement is to the true value.

Repeatable and reproducible data

Each time a test is carried out the same method is used. This means that if someone tested the same rope twice they would expect to get the same result – the measurement is **repeatable**. It also means that if anyone else carried out the test using this method they would expect to get the same results – the measurement is **reproducible**.

In reality there may be small differences in the results obtained each time a test is carried out. This is because there may be errors, for example, errors within the machine. These errors can be kept small by regularly calibrating the machine. Calibration is used to check that the readings given by the machine are accurate and adjusting the machine if they are not.

It is important to control all the factors that are not being tested that may affect the results, for example, temperature. The minimum breaking load is the force needed to break the rope at standard room temperature. If the rope was tested at a different temperature this might affect the strength of the rope and you could not be sure that the rope meets the requirements.

The best estimate of the true value of the breaking load is found by repeating measurements on at least three samples, then calculating the mean and finding the range.

Abseiling ropes must be strong in tension. Climbers want to be sure that their ropes have been tested.

Key words

✓ **accuracy**
✓ **repeatable**
✓ **reproducible**

Questions

4 Give one factor that is controlled when abseiling ropes are tested.

5 Explain why is it important to repeat measurements.

6 Samples of abseiling rope were tested for their strength in tension and the following results were collected:

27 546 N   27 356 N   27 598 N   27 467 N.

Calculate the mean and the range for this set of measurements.

Find out about

- ✔ **materials under the microscope**
- ✔ **molecules and atoms in materials**
- ✔ **models of molecules**

A woollen jumper is very different from a silk shirt. The shirt is more formal and less stretchy than the jumper. They are both made from natural polymers but they are very different. Their properties depend on their make-up, from the large scale to the invisibly small:

- the visible weave of a fabric
- the microscopic shape and texture of the **fibres**
- the molecules that make up the polymer
- the atoms that make up the molecules.

The visible weave

The fabric of a woven shirt is tightly woven but even so it is possible to see the criss-cross pattern of threads. The fabric is hard to stretch because the strong threads are held together so tightly.

On the other hand, a knitted jumper is soft and stretchy. The loose stitches allow the threads to move around.

The weave and the stitches are visible to the naked eye. They are **macroscopic** features. However, the properties of a fabric also depend on smaller structures.

Magnification: × 20. Visible: to naked eye. Width of circle: 4 millimetres.

Magnification: × 1000. Visible: down a microscope. Width of circle: 80 micrometres.

Magnification: × 50 million. Visible: not even in a microscope. Width of circle: 1.5 **nanometres**.

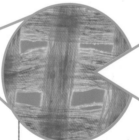  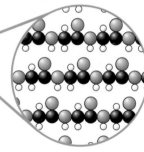

Levels of structure and detail. A millimetre is a thousandth of a metre. A micrometre is a thousandth of a millimetre. And a nanometre is a thousandth of a micrometre.

Taking a closer look

A microscope can show details of the individual fibres in a fabric. Silk, for example, has smooth, straight fibres that slide across each other.

Wool fibres have a rough surface that is covered in scales. The wool fibres tend to cling to each other in the thread and also make the threads cling together.

Silk.

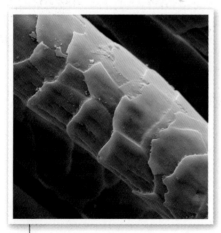

A wool fibre magnified 1250 times.

The invisible world of molecules

It is difficult to look much further into the structure of materials using microscopes of any kind. Scientists explain the differences between silk, wool, and other fibres by finding out about their molecules. Molecules are very small indeed, so small that it needs a giant leap of the imagination to think about them.

Scientists measure the sizes of atoms and molecules in nanometres (nm). There are 1 000 000 000 nanometres in a metre. Some molecules, such as the small molecules in air, are even smaller than one nanometre but many are bigger.

The molecules in fibres are big on the nanometre scale. They are very long – 1000 nanometres or more. The shape and size of the **long-chain molecules** in a fibre make the material what it is. Polymers have special properties because the molecules in them are so long.

Model molecules

Even the largest molecules and atoms are invisible. So in the nanoworld of molecules, scientists build models based on the results from their experiments.

Models of molecules can be compared with the map of the London tube system. The tube map does not look like an underground railway. But it has lots of useful information about the way the stations are connected. In a similar way, models of molecules do not look like real molecules. But they show what scientists have discovered about the atoms in the molecules and how they are joined together.

Computer model of a protein molecule. No-one knows what atoms and molecules look like. It helps to use models to understand what they do. In the real world the atoms are not coloured. In this computer image the atoms of each element are colour-coded: carbon (green), sulfur (yellow), nitrogen (blue), hydrogen (grey), and oxygen (red).

Questions

1. a Put the following in order of size, starting with the largest: fibre, fabric, atom, thread, molecule.
   b Use the words in part **a** to write four sentences that describe the decreasing structures. The first sentence might be: Fabrics are made by weaving together threads.

2. a How many chemical elements are there in silk?
   b A hydrocarbon contains hydrogen and carbon atoms only. Is silk a hydrocarbon?

3. A polymer molecule is about 1000 nanometres long. An atom is about 0.1 nanometres across.
   a Estimate how many atoms there are along the chain.
   b How many molecules would fit into a millimetre?

Find out about

- polymer discoveries
- polymers as long-chain molecules
- monomers that make up polymers

The 1930s was the decade of the first synthetic polymers. The world was a tense place and war was on its way. Governments were looking for scientific solutions to give them an advantage. This speeded up many scientific developments. Some of these used the big new idea: polymers. However, the first synthetic polymer was discovered by accident.

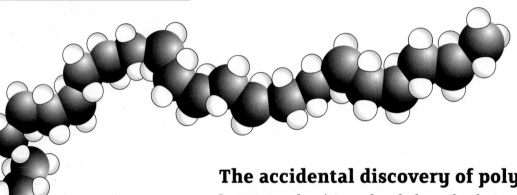

The accidental discovery of polythene

In 1933, two chemists made polythene thanks to a leaky container. Eric Fawcett and Reginald Gibson were working for ICI. Their job was to investigate the reactions of gases at very high pressures. They had put some ethene gas into the container and squashed it to 2000 times its normal pressure. However, some of the ethene escaped. When they added more ethene, they also let in some air.

Two days later, they found a white, waxy solid inside the apparatus. This was a surprise. They decided that the gas must have reacted with itself to form a solid. They realised that, in some way, the small molecules of ethene had joined with each other to make bigger molecules.

They worked out that the new molecules were like repeating chains. The chains were made from repeating links of ethene molecules.

Later they understood that oxygen in the air leaking into their apparatus had acted as a catalyst. The oxygen speeded up what would otherwise have been a very, very slow reaction to join the ethene molecules together.

What are polymers?

Polymers all have one thing in common: their molecules are long chains of repeating links. Each link in the chain is a smaller molecule. These small molecules are called **monomers**. They each connect to the next one to form the chain. This is true for natural polymers such as cotton, silk, and wool and for synthetic polymers such as polythene, nylon, and neoprene.

The common name for the polymer discovered by Fawcett and Gibson is polythene. This is short for the chemical name poly(ethene). The word poly means 'many'; a poly-ethene molecule is made from many ethene molecules joined together.

A polymer pioneer

Wallace Carothers was an American chemist who discovered neoprene and invented nylon. Neoprene was another accidental discovery. A worker in Carothers' laboratory left a mixture of chemicals in a jar for five weeks. When Carothers had a tidy up, he discovered a rubbery solid in the bottom of the jar. Carothers realised that this new stuff could be useful. He developed it into neoprene. This synthetic rubber first came on the market in 1931 and is still used today, to make wetsuits, for example. This discovery helped Carothers to work out a theory of how small molecules can **polymerise**.

The discovery of nylon

Japan and the USA were on bad terms in the years before World War II. Trade between them was difficult and the supply of silk was cut off. It became rare and expensive. Carothers started looking for a synthetic replacement. In 1934, his team came up with nylon. This is a polymer made from two different monomers. The different molecules join together as alternate links in the chain.

Sadly, Carothers died before he could see the effects of his discoveries. Nevertheless, they are both still in use today.

ethene gas under pressure

The original high-pressure container used by Fawcett and Gibson is on display at the Science Museum. The diagrams show what was happening to the small ethene molecules as they joined up in long chains to make polythene. This is polymerisation.

Key words
- monomer
- polymerise

Questions

1 What is a polymer, and what is a monomer?

2 a Write down the names of two polymers that were discovered by accident.
   b Many scientists have made accidental discoveries. All of these words might be used to describe these scientists:

lucky, skilful, foresight, inventive, creative

Choose two of these words to describe the scientists. In each case, explain why you have chosen that word.

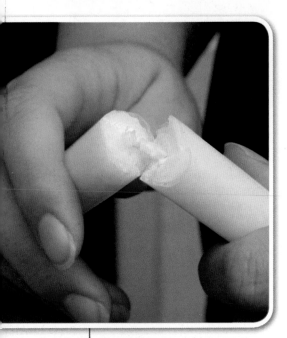

The molecules of candle wax are about 20 atoms long. Wax is weak and brittle.

The molecules of polythene are similar to those of candle wax. But they are about 5000 times longer. Polythene is much stronger and tougher than candle wax.

Long and short molecules

The properties of a polymer depend on the length of its molecules. The molecules in candle wax are very similar to those in polythene. However, wax is weaker and more brittle than polythene. Wax also melts at a lower temperature. This is because the wax molecules are much shorter. They contain only a few atoms; polythene molecules contain many thousands.

Two different bonds

The molecules are made of atoms. The bonds between *atoms* in the molecules are strong. So it is very hard to pull a molecule apart. The molecules do not break when materials are pulled apart.

But the forces between *molecules* are very weak. It is much easier to separate molecules from one another. They can slide past each other.

Breaking and melting wax and polythene

Stretch or bend a candle and it cracks. This is because separating the small molecules is not difficult. The forces between the molecules are very weak and they slip past each other quite easily.

Breaking a lump of polythene is much more difficult. Its long molecules are all jumbled up and tangled. It is harder to make them slide over each other. The long molecules make polythene stronger than wax.

Polythene also has a higher melting point than wax. This is because the forces between long polythene molecules are slightly stronger than the forces between short wax molecules. More energy is needed to separate the polythene molecules from each other so it melts at a higher temperature.

Question

1 Bowls of pasta can be used as an analogy to explain the difference between wax and polythene. One bowl contains cooked spaghetti. The other bowl contains cooked macaroni (or penne).
   a In the analogy, what represents a molecule?
   b Which kind of pasta represents wax and which represents polythene?
   c Show how this analogy can help to explain why polythene is stronger than wax.

Designer stuff

Hardening rubber

Natural rubber is a very flexible polymer, but it wears away easily. This makes it good at rubbing away pencil marks, but not much else.

Sometime around 1840, an American inventor called Charles Goodyear was experimenting with mixing sulfur and rubber. He was trying to improve the properties of the natural material. He accidentally dropped some of his mixture on top of a hot stove. He didn't bother to clean it off, and the next morning it had hardened.

It took two more years of research to find the best conditions for this new process, which Goodyear called **vulcanisation**. It made rubber into a stronger material that was more resistant to heat and wear. At that time no-one knew why this happened. They just knew that it worked and that it made rubber an excellent material for car tyres.

Goodyear started a business making tyres. He began with tyres for bicycles and prams. Now the business makes tyres for cars, motorcycles, and aeroplanes.

Cross-links

Goodyear was the first person to alter the properties of a polymer. He did not know why vulcanisation worked – only that it did. Now that we understand more about molecules, we know what's going on.

The sulfur makes **cross-links** between the long rubber molecules. The molecules are locked into a regular arrangement. This stops them from slipping over each other and makes the rubber less flexible, stronger, and harder. It also gives the rubber a higher melting point because more energy is needed for the molecules to separate and break out of the solid.

Find out about

- using science to change polymer properties
- cross-links to make polymers harder
- plasticisers to make polymers softer
- polymers that are crystalline

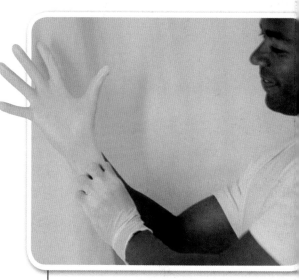

Vulcanising natural rubber produces gloves that are strong enough not to tear.

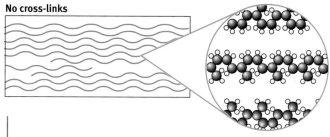

No cross-links

Each line represents a polymer molecule. Without cross-linking, the long chains can move easily, uncoil, and slide past each other.

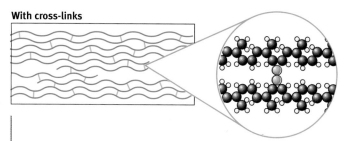

With cross-links

The sulfur atoms form cross-links across the polymer chains. This stops the rubber molecules uncoiling and sliding past each other.

Softening up

PVC is a polymer often used for making window frames and guttering. These need to be **durable** and hard. PVC is also a good polymer for making clothing, but for this purpose it needs to be softer and more flexible.

To make PVC softer, the manufacturer adds a **plasticiser**. This is usually an oily liquid with small molecules. The small molecules sit between the polymer chains.

The polymer chains are now further apart. This weakens the forces between them, so less energy is needed to separate them and they slide over each other more easily. This means the polymer is softer and more flexible, and has a lower melting point.

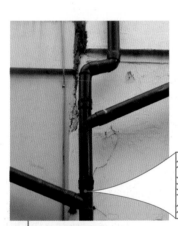

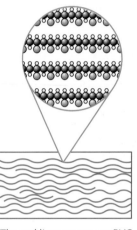

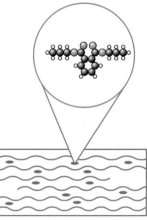

This PVC is unplasticised. It is called uPVC.

The red lines represent PVC molecules. These PVC molecules are long chains that lie close together. The closer they are, the stronger the forces between them.

This PVC has been plasticised to make it soft.

The molecules of plasticiser hold the PVC chains apart. This weakens their attraction and makes it easier for them to slide past each other.

Question

1 a Chemists can vary the extent of cross-linking between chains in rubber. How would you expect the properties of rubber to vary as the degree of cross-linking increases?

b What effect do plasticisers have on the properties of polymers?

Cling film

Cling film was first made from plasticised PVC. Unfortunately, the small plasticiser molecules were able to move through the polymer and into the food. Some people worried that the plasticiser might be bad for their health. The evidence that plasticisers are harmful is controversial and strongly challenged by the plastics industry (see C3, K: Benefits and risks of plasticisers).

There is stronger evidence that the regular use of plastic food wrap can cut down on food poisoning, which is a serious and growing risk to health.

Some cling film is now made using PVC and plasticisers that are much less likely to move from the polymer to food.

Original polythene

Chemists now know that in polythene made under high pressure the polymer chains have branches, which stop the molecules packing together neatly.

Compare a bonfire pile with a log pile. In a bonfire pile, the twigs and side branches stick out all over the place and the pile is full of holes. But a pile of logs is neatly stacked. It is the same with polymers but on a much smaller scale. If there are side chains, the structure is messy and full of holes. This is the case with polythene made under pressure, which has **branched chains**.

A stronger, denser polythene

In some polymers the molecules are all jumbled up in a very irregular arrangement. In other polymers the molecules are arranged in neat lines; these are called **crystalline polymers**. There are also polymers that are partly irregular and partly crystalline. Scientists realised that increasing the crystallinity of polythene might make it stronger and denser. In the 1950s scientists found a way of making polythene molecules in neat piles of straight lines. It was an international effort by a German, Karl Ziegler, and an Italian, Giulio Natta.

They used special metal compounds as catalysts. These metal compounds act in a similar way to the oxygen in the high-pressure process. They speed up the rate at which the ethene molecules join together. The growing polymer chains latch onto the solid catalyst. The regular surface of the solid allows the molecules to build up more regularly.

In this new crystalline form of polythene, the molecules are more neatly packed together. The forces between the molecules are slightly stronger and more energy is needed to separate them. The new form is stronger and denser than the older type and softens at a higher temperature. Both types are still made – the old branched-molecule form is called low-density polythene (LDPE). The newer crystalline form is high-density polythene (HDPE).

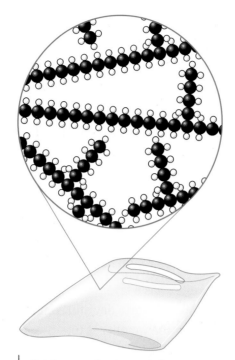

Polythene molecules made from ethene under pressure have side branches. This stops the polymer molecules lining up neatly. This type of polythene has a slightly lower density and is not crystalline.

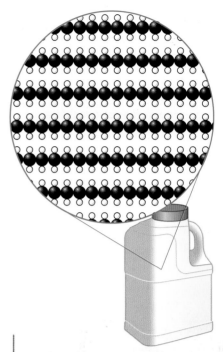

Polythene molecules made from ethene with a special catalyst do not have side branches. The polymer molecules line up neatly. This type of polythene has a slightly higher density and is crystalline.

Key words

✓ **branched chains**
✓ **crystalline polymers**

Questions

2 Why is HDPE slightly denser than LDPE? Suggest an explanation based on the structure and arrangements of molecules.

3 LDPE starts to soften at the temperature of boiling water. HDPE keeps its strength at 100°C. Suggest some products that would be better made of HDPE rather than LDPE and give your reasons.

Find out about

✔ **using science to design new polymer products**

Gore-tex is waterproof and windproof, yet it allows the moisture from sweat to pass through.

Key word

✔ **melting point**

Ingenious layers: Gore-tex

Sometimes layers of different polymers with different properties are sandwiched together. An interesting example is the waterproof fabric Gore-tex, named after its inventor Bob Gore. He was working with a polymer called PTFE. This is the plastic coating for non-stick pans.

Gore discovered that if a sheet of PTFE is stretched, it develops very small holes and becomes porous. A single water molecule can pass through the small holes. But a whole water droplet is too large to get through. This got Gore thinking. His idea was that vapour evaporating from someone's skin would pass through the polymer sheet, but that rain drops would not.

Gore-tex has a layer of PTFE sandwiched between two layers of cloth. The wearer stays dry and comfortable no matter how energetic they are or what the weather is like. Sweat can always pass out through the fabric, but no water can get in.

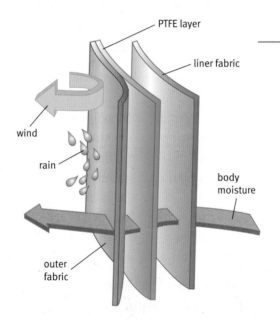

Gore-Tex membrane: there are billions of tiny holes in the film of PTFE. The holes are 20 000 times smaller than a raindrop but 700 times larger than a water molecule.

A very strong polymer: Kevlar

Nearly all the early synthetic polymers were discovered by accident. But once chemists started to understand how polymerisation works, they could predict how reactions might take place. This meant they could plan to make a polymer with certain properties.

Du Pont is a huge multinational company with a special interest in polymers. The company wanted to make a very strong but light-weight polymer with a high **melting point**. The chemists designed and made a polymer with very long molecules, linked together in sheets. These sheets were themselves tightly packed together in a circular pattern.

One of the scientists involved in the research was an American, Stephanie Kwolek. Her job was to make small quantities of the new polymer and turn it into a liquid. Once it was liquid, the polymer could be forced through a small hole to make fibres. The problem was that the polymer would not melt, nor would it dissolve in any of the usual solvents.

Stephanie Kwolek experimented with many solvents. She eventually found that the new polymer would dissolve in concentrated sulfuric acid. This is a highly dangerous chemical that can cause severe burns. But fibres of the new polymer were manufactured in this way. This was the origin of Kevlar, which is five times stronger than steel. It is used for bullet-proof vests and to reinforce tyres. A similar polymer called Nomex is used in protective clothing for racing drivers.

Copying nature: Velcro

The two surfaces of Velcro stick together with a strong bond but can be peeled apart. One surface is covered in hooks, the other in loops.

The inventor of Velcro, George de Mestral, was copying seed pods that he found stuck to his socks when he was out walking. The pods were covered with tiny hooks that attached themselves round threads in the socks.

De Mestral used nylon to make Velcro. He worked out how to weave the polymer thread in just the right way to produce hooks and loops.

Stephanie Kwolek wearing protective gloves made of Kevlar. She discovered how to turn this polymer into fibres.

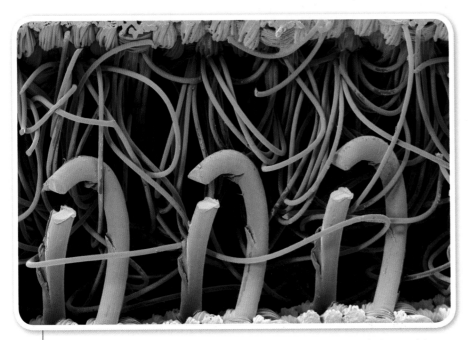

A magnified view of the nylon hooks and loops in Velcro material. This is a false-colour image taken with an electron microscope. The loops are loosely woven strands. The hooks are loops woven into the fabric and then cut. When the two surfaces are brought together they form a strong bond, which can be peeled apart. Magnification × 30.

Questions

1 Look through Sections E–H and identify two examples each of polymers, or polymer products:
   a discovered by accident
   b developed by design.

2 Explain how Gore-tex is waterproof, but still allows water to pass through it.

3 What words describe the properties of the polymer needed to make:
   a the hooks in Velcro?
   b the loops in Velcro?

Plant for processing chemicals from oil.

A fractionating tower, where crude oil is separated into fractions.

Crude oil

Polymers are made from small molecules called monomers joined together in long chains. In most synthetic polymers, the small molecules originally come from **crude oil**.

Crude oil is a thick, sticky, dark-coloured liquid that formed over millions of years from the remains of tiny plants and animals called plankton. It is pumped out of the Earth's crust from wells under the ground or sea.

Crude oil is not very useful as it is.

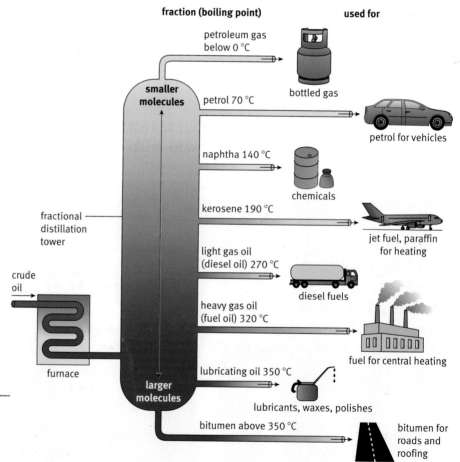

fraction (boiling point) — used for

petroleum gas below 0 °C — bottled gas

smaller molecules

petrol 70 °C — petrol for vehicles

naphtha 140 °C — chemicals

fractional distillation tower

kerosene 190 °C — jet fuel, paraffin for heating

light gas oil (diesel oil) 270 °C — diesel fuels

crude oil

heavy gas oil (fuel oil) 320 °C — fuel for central heating

furnace

lubricating oil 350 °C — lubricants, waxes, polishes

larger molecules

bitumen above 350 °C — bitumen for roads and roofing

Crude oil is a mixture of **hydrocarbon** compounds. Because it is a mixture, crude oil is not very useful as it is. It needs to be separated into groups of molecules of similar size, called **fractions**. This is done by **fractional distillation**. When crude oil has been refined in this way it becomes very useful indeed, which is why it is so valuable.

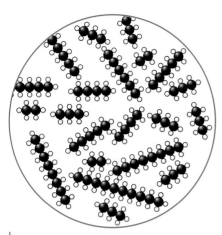

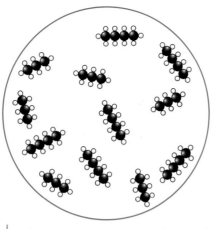

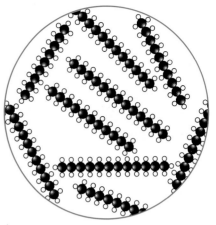

Crude oil is a mixture of hundreds of different hydrocarbons. A hydrocarbon contains hydrogen and carbon atoms only.

These molecules go right to the top of the fractionating column. This fraction contains some of the shortest hydrocarbons. There are weak forces between the molecules, giving them a low boiling point.

As you go down the tower the fractions have molecules with longer hydrocarbon chains. The forces between them get stronger, which increases the boiling points.

In the diagram on page 154, the crude oil is being heated in a furnace. The hydrocarbons in crude oil then go into the fractionating tower, which is hottest at the bottom and coolest at the top. The hydrocarbon molecules are separated by their boiling points.

The smallest molecules have the lowest boiling points and go to the top of the tower. This is because the forces between these molecules are very weak and only a little energy is needed for them to break out of the liquid and form a gas. The biggest molecules have the highest boiling points because the forces between the molecules are slightly stronger. These molecules stay at the bottom of the tower. Each fraction produced is still a mixture of molecules, but they are of similar size and boiling point.

Each fraction has different uses related to its properties. Most of the fractions from crude oil are used to provide energy for transport, homes, and industry. Some are used as lubricants. Some of the larger molecules are broken into smaller pieces, which can be used in **chemical synthesis** to manufacture new materials such as polymers. Only about 4% of crude oil is used in this way.

Key words

- ✓ crude oil
- ✓ hydrocarbon
- ✓ fraction
- ✓ fractional distillation
- ✓ chemical synthesis

Questions

1 Crude oil is made of hydrocarbons.
  a What is a hydrocarbon?
  b How do the hydrocarbons in crude oil vary?

2 Copy and complete:

  The fractionating tower separates the hydrocarbons into groups of molecules called _____. These groups of molecules are still mixtures, but they contain a _____ number of different hydrocarbons than the original crude oil. The hydrocarbon molecules in a fraction are similar in _____.

3 Explain why the fractional distillation of crude oil is so important.

Find out about

- ✔ the size of a nanometre
- ✔ the properties of nanoparticles
- ✔ new materials containing nanoparticles

Some bus companies use a nanoscale additive in diesel fuel. This reduces the amount of fuel used and the emissions from the vehicle, making the buses more efficient.

Geckos' feet have millions of nanometre-sized hairs that provide the 'stickiness' it needs to walk on ceilings.

Not just small – very small

Bus companies, holidaymakers, and people with serious injuries are just some of the groups benefiting from recent scientific and technological advances. New products have been developed that rely on the properties of materials at a very, very small scale. These technologies are called **nanotechnology**. 'Nano' comes from the Greek work *nanos*, which means dwarf. The particles used in nanotechnology are measured in nanometres (nm).

Nanotechnology is the use and control of structures called **nanoparticles**, which are very small. Some nanoparticles are made using specialist tools that build up new structures atom by atom. Other nanoparticles are made by chemical synthesis or by other techniques such as etching.

Each nanometre (nm) is a billionth of a metre or 0.000 000 001 m. A nanometre is about:

- the width of a DNA molecule
- the distance your fingernails grow in a second
- 1/80 000 the thickness of the average human hair.

It is difficult to understand just how tiny a nanometre is.

The diagram on the right shows a nanoparticle. It looks like a tiny piece of soot. It is 10 nm across, which is about 0.00001 mm. You could fit about 100 000 of them across 1 mm on your ruler. Let's think about it another way.

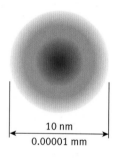

10 nm
0.00001 mm

Imagine if a nanoparticle about 10 nm across was scaled up to the size of a football.

- An atom would become about the size of a 10 p coin.
- A red blood cell would be the size of a football pitch.
- A cat would be about the same size as the Earth.

Properties of nanoparticles

Size of forces

If you wear a jumper and shirt both made of synthetic polymers, they sometimes seem to stick together. The force causing this 'stickiness' is similar to the force that allows a balloon rubbed against a jumper to stick to a wall. Usually we don't notice the forces holding surfaces together because they are very weak, but on the nanoscale they become very strong. These forces have the potential to be useful; they are what allow geckos to walk on ceilings. The geckos' feet have millions of tiny, nanometre-sized hairs, which give it a huge surface area and provide the stickiness required to hold the gecko to the ceiling. This has inspired scientists in the USA to design a robot that can walk up walls. They use a polymer that is similar to the bottom of the geckos' feet.

Size of surface

For nanoparticles, the **surface-area**-to-volume ratio is very large. In a solid 30-nm particle, about 5% of the atoms are on the surface. In a solid 3-nm particle, about half are. The atoms on the surface tend to be more reactive than those in the centre. This means materials containing nanoparticles are often highly reactive or have unusual properties.

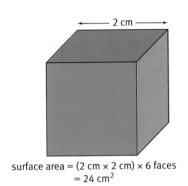
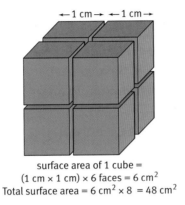

surface area = (2 cm × 2 cm) × 6 faces
= 24 cm^2

surface area of 1 cube =
(1 cm × 1 cm) × 6 faces = 6 cm^2
Total surface area = 6 cm^2 × 8 = 48 cm^2

Total surface area is larger for smaller particles.

Nanotechnology in medicine

Silver is best known for use in jewellery because it is shiny and unreactive. For a long time, it has been known that silver also has antibacterial properties. In 1999, a new bandage for serious injuries was launched. It is called Acticoat and contains silver nanoparticles. It is particularly important that the bacteria in a serious wound are killed before they can cause infections. The tiny nanoparticles of silver dissolve very quickly once they are moistened (for example, by blood from the wound) and the silver can get to work straight away.

Some sunscreens contain nanoparticles.

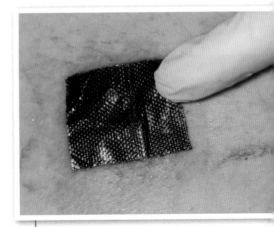

Acticoat is a type of bandage that contains nanometre-sized silver particles with antibacterial properties.

Questions

1 Write a description of what a nanometre is, for a student in Year 7.

2 Silver has many different uses.
   a Which properties of silver make it good for making jewellery?
   b How are the particles of silver in Acticoat different from the silver particles in jewellery?
   c Why will nanoparticles of silver be more effective in a wound dressing than ordinary silver?

Key words

✓ **nanotechnology**
✓ **nanoparticles**
✓ **surface area**

Nanotechnology in nature

Nanotechnology sounds very strange and new – but there are nanoparticles in nature. There are even nanoscale structures in living cells that can move and turn in a controlled way and carry out complex jobs. They are far more advanced than any of the synthetic nanotechnologies currently in use.

All these are on the nanoscale:

- tiny salt particles in the atmosphere, formed by ocean waves in windy conditions, which help in forming rain and snow
- proteins that control biological systems very precisely
- the enamel in your teeth, which is partly made of nanoparticles.

Humans have been making nanoparticles for years by accident, without knowing it. Some fires, particularly those burning solid fuels, produce nanoparticles (along with other waste).

Uses of nanotechnology

There are products already available that use nanotechnology. These include healthcare products, sports gear, and clothing.

Sunscreen

Many sunscreens contain particles of either zinc oxide or titanium oxide. These are white solids. In older formulations, the particles are relatively large and leave the skin looking white. More modern formulations use nanoparticles instead, which can be rubbed in and have a more natural appearance.

Tennis balls and rackets

Nanotechnology was first used in a professional tennis match in 2002. The 'double-core' balls have an extra layer inside them made of 1-nm-sized clay particles mixed with rubber. This helps to slow down air escaping from the balls, keeping them inflated for longer.

Nanoparticles are also added to materials such as the carbon fibre used to make tennis rackets. The resulting materials are lighter and stronger.

Nanoparticles of salt are formed above the sea.

Older sunscreens could be seen on the skin.

Nanotechnology is improving the performance of sports equipment.

Electronic paper

Engineers at Bridgestone have developed a flexible display using nanotechnology. A substance they call 'liquid powder' is placed between two sheets of glass or plastic, creating a light-weight, flexible display. Electricity is needed to change the display, but when it is switched off it retains its image. These displays could replace paper posters, saving paper – and electric signs, saving energy.

Clothing

Scientists have developed clothes that contain zinc oxide nanoparticles. These are the same particles as those used in sunscreen. Clothes with these particles offer better UV protection. Stain-resistant clothes have also been produced. These have tiny nanoscale hairs that help repel water and other materials. This has the potential to reduce the amount of water and energy used in washing clothes.

Socks have been made that contain nanoparticles of silver or other chemicals. This gives the socks antibacterial properties to help prevent feet from smelling.

Coatings containing nanoparticles can repel water and other chemicals that might stain.

Self-cleaning windows

A company called Pilkington offers a product called Activ Glass, which is coated in nanoparticles. When light hits these particles, they break down any dirt on the glass. The surface is also hydrophilic (*hydro* - water, *philic* - loving), which means that water falling on it spreads over the surface, helping to wash it.

Nanotechnology and risk

Different properties, different risks

Nanoparticles have different properties compared to larger particles of the same material. This may mean that the nanoparticles have different effects on plants, animals, and the environment. It may also mean that they are more toxic to people. Some doctors are concerned that nanoparticles are so small they may be able to enter the brain from the bloodstream. If this is true, it could mean some chemicals that are normally harmless become highly toxic at the nanoscale.

Exactly how all the various nanoscale substances differ from larger particles of the same material is not fully understood. At present, there are no requirements for health and safety studies for nanoparticles to be different from those for larger particles. But some groups and organisations think that there should be.

Questions

1 Using Sections J and K, give an example of a sports product, a healthcare product, a lifestyle product, a building material, and clothing that may now contain nanotechnology.

2 List three ways in which nanotechnology products may be beneficial to the environment.

3 Explain why some people are concerned about the possible effects of nanotechnology on the environment.

Science Explanations

Scientists use their knowledge of molecules to develop new materials with useful properties. A wide range of different synthetic polymers can be made from hydrocarbons obtained from crude oil.

You should know:

- that one way of comparing materials is to measure their properties, such as melting point, strength, stiffness, hardness, and density
- why it helps to have an accurate knowledge of the properties of materials when choosing a material for a particular purpose
- that polymers such as plastics, rubbers, and fibres are made up of long-chain molecules
- why modern materials made of synthetic polymers have often replaced materials used in the past such as wood, iron, and glass
- that crude oil is one of the raw materials from the Earth's crust, which is used to make synthetic polymers
- that crude oil consists mainly of hydrocarbons, which are chain molecules of varying lengths
- that hydrocarbons are made from carbon and hydrogen atoms only
- why the boiling temperature of a hydrocarbon depends on its chain length
- that the petrochemical industry makes useful products by refining crude oil
- that fractional distillation separates the hydrocarbons in crude oil into fractions according to their chain length
- that some of the small molecules from refining crude oil are used to make new chemicals
- that polymerisation is a chemical reaction that joins up small monomer molecules into long chains
- how the properties of polymeric materials depend on the way in which the long molecules are arranged and held together
- how it is possible to modify polymer properties in various ways such as increasing the length of the chains, cross-linking the molecules, adding plasticisers, and changing the degree of crystallinity
- that nanotechnology is the use and control of structures that are very small (1–100 nm)
- that nanoparticles can occur naturally, by accident, and by design
- why nanoparticles of a material show different properties to larger particles of the same material
- how nanoparticles can be used to modify the properties of materials.

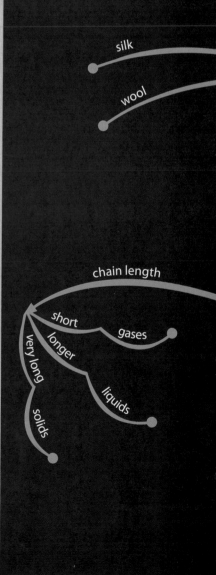

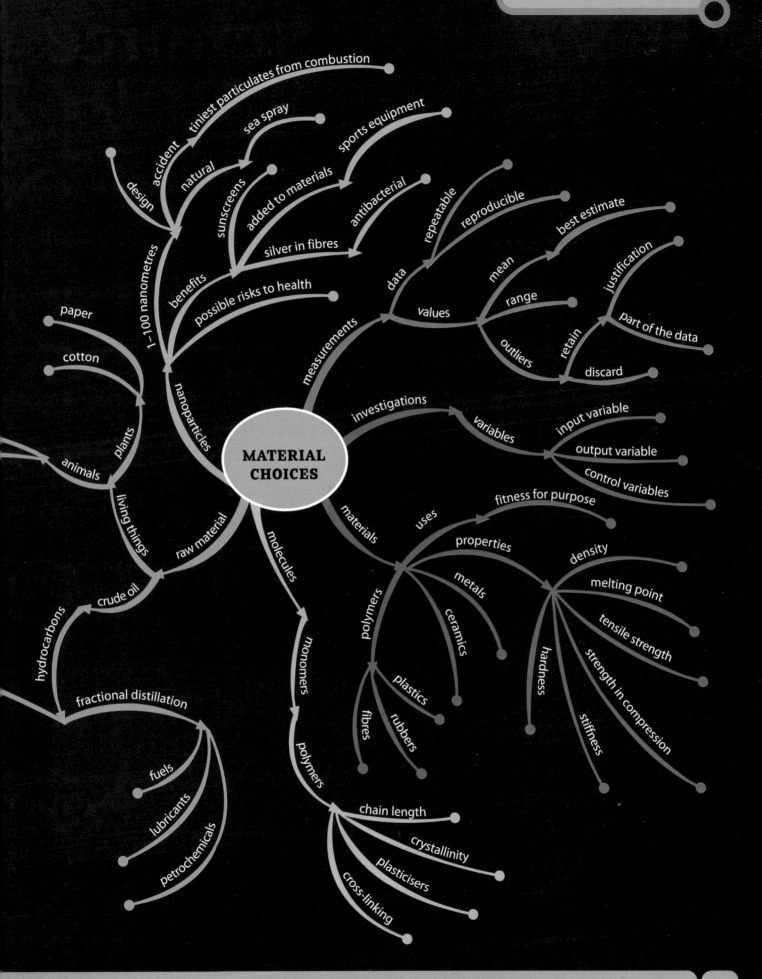

MATERIAL CHOICES

nanoparticles
- 1–100 nanometres
- design
- accident
- tiniest particulates from combustion
- natural
  - sea spray
- sunscreens
- added to materials
  - sports equipment
  - antibacterial
- silver in fibres
- benefits
- possible risks to health

measurements
- data
  - repeatable
  - reproducible
- values
  - mean
    - best estimate
    - justification
  - range
  - outliers
    - retain
      - part of the data
    - discard

investigations
- variables
  - input variable
  - output variable
  - control variables

raw material
- living things
  - plants
    - paper
    - cotton
  - animals
- crude oil
  - hydrocarbons
  - fractional distillation
    - fuels
    - lubricants
    - petrochemicals

molecules
- monomers
- polymers
  - chain length
  - crystallinity
  - plasticisers
  - cross-linking

materials
- uses
  - fitness for purpose
- properties
  - density
  - melting point
  - tensile strength
  - strength in compression
  - stiffness
  - hardness
- metals
- ceramics
- polymers
  - plastics
  - rubbers
  - fibres

Ideas about Science

Scientists measure the properties of materials to decide what jobs they can be used for. Scientists use data rather than opinion to justify the choice of a material for a purpose.

Scientists can never be sure that a measurement tells them the true value of the quantity being measured. Data is more reliable if it can be repeated. When making several measurements of the same quantity, the results are likely to vary. This may be because:

* you have to measure several individual examples, for example, several samples of the same material
* the quantity you are measuring is varying, for example, different batches of a polymer made at different times
* the limitations of the measuring equipment or because of the way you use the equipment.

Usually the best estimate of the true value of a quantity is the mean (or average) of several repeat measurements. The spread of values in a set of repeat measurements, the lowest to the highest, gives a rough estimate of the range within which the true value probably lies. You should:

* be able to calculate the mean from a set of repeat measurements
* know that a measurement may be an outlier if it lies well outside the range of the other values in a set of repeat measurements
* be able to decide whether or not an outlier should be retained as part of the data or rejected when calculating the mean.

When comparing data on the properties of different samples of a material you should know that:

* a difference between their means is real if their ranges do not overlap.

To investigate the relationship between a factor and an outcome, it is important to control all the other factors that you think might affect the outcome. In a plan for an investigation into the properties of a material, you should be able to:

* identify the effect of a factor on an outcome
* explain why it is necessary to control all the factors that might affect the outcome other than the one being investigated
* recognise that the control of other factors is a positive design feature or that it is a design flaw if they are not controlled.

Some applications of science, such as using nanoparticles, can have unintended and undesirable impacts on the quality of life or the environment. Benefits need to be weighed against costs. You should know that:

* some nanoparticles may have harmful effects on health and that there is concern that products containing nanoparticles are being introduced before these effects have been fully investigated.

Review Questions

1 A company tested the minimum breaking strength (kN) of a 12-mm-diameter nylon rope and a 14-mm-diameter polypropylene rope. The test results are in the table.

Test number	1	2	3	4	5
Nylon rope	25.7	25.1	24.8	25.2	25.2
Polypropylene rope	19.9	20.8	20.1	20.4	20.8

  a Suggest why the company repeated the test on each rope five times.

  b For each rope, calculate the best estimate of the true value of the strength of the rope.

  c Is there a real difference between the strengths of the two ropes? Use the test results to justify your answer.

2 **a** The table gives data about three hydrocarbons. The higher the number of carbon atoms in a molecule, the bigger the molecule. Describe the trend in boiling point as molecule size increases.

Hydrocarbon name	Boiling point (°C)	Number of carbon atoms in one molecule
Methane	−162	1
Hexane	69	6
Hexadecane	287	16

  b Explain the trend in boiling points in terms of the strength of the forces between the molecules.

3 **a** Identify the property changes that occur when cross-links are made between long rubber molecules. Choose from the list below.
  - flexibility increases
  - flexibility decreases
  - hardness increases
  - hardness decreases
  - melting point increases
  - melting point decreases

  b Identify two property changes that occur when a plasticiser is added to a polymer.

4 **a** Titanium dioxide nanoparticles are used in some sunblock creams. Because they are so small they do not reflect visible light.
  i Within what range is the diameter of a nanoparticle?
  ii Suggest one advantage of using nanoparticles in sunblock creams.
  iii Explain why some people are concerned about the safety of using sunblock creams that contain nanoparticles.

  b Give examples of two more uses of nanoparticles.

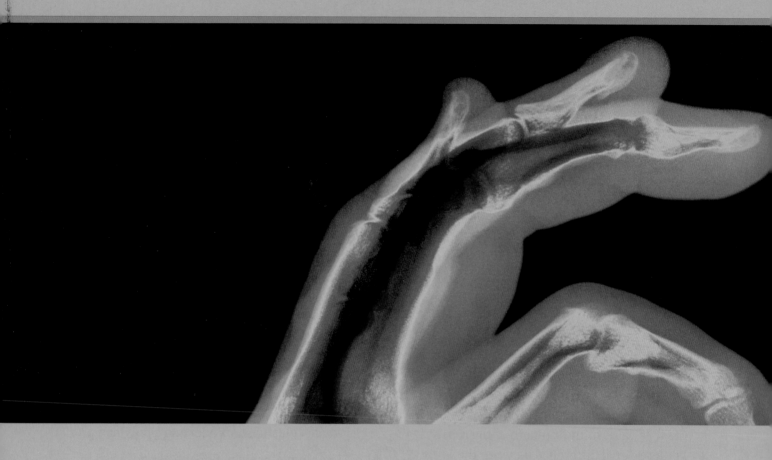
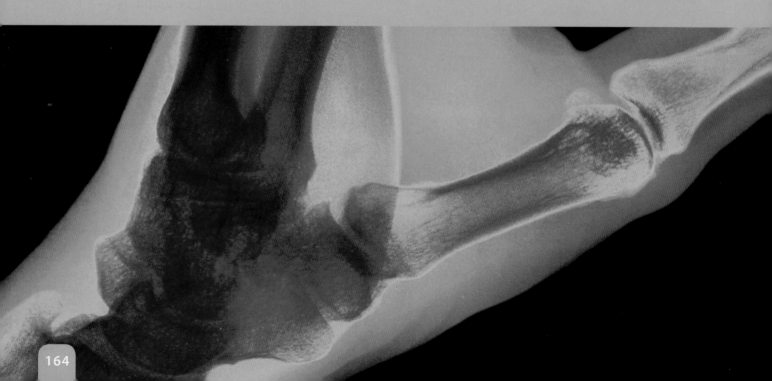

P2 Radiation and life

Why study radiation and life?

Human eyes detect one type of radiation – visible light. But there are many other types of 'invisible' radiation. Radiation can be harmful. You hear a lot about the health risks of different radiations. For example, from natural sources such as sunlight, and from devices such as mobile phones. Radiation is involved in climate change, and this is the biggest risk of all.

What you already know

- Light travels in straight lines.
- Light travels very fast.
- When light strikes an object, some may be reflected, some absorbed, and some passes straight through.
- White light can be split into the colours of the spectrum.

Find out about

- how radiation travels and is absorbed
- the evidence of global warming, and its effects
- how information is stored and transmitted digitally
- microwave radiation from mobile phones
- weighing up risks and benefits.

The Science

There are several types of radiation that belong to one 'family', the electromagnetic spectrum. Our communications systems use electromagnetic radiation for transmitting information, such as radio and TV programmes, mobile phones, and computer networks. Science can explain what happens when radiation is absorbed by our bodies. It can also explain how radiation warms the atmosphere, and uses computer modelling to predict global warming.

Ideas about Science

To make sense of media stories about radiation, you need to understand a few things about correlation and cause. How sure can we be that one thing causes another? You can also learn how to evaluate reports from health studies, and how to interpret statements about risk.

Find out about

- ✔ **benefits and risks of exposure to sunlight**
- ✔ **the electromagnetic spectrum**

Skin colour

The **ultraviolet radiation (UV)** in sunlight can cause skin cancer. Skin cancer can kill.

Melanin, a brown pigment in skin, provides some protection from UV radiation. People whose ancestors lived in sunnier parts of the world are more likely to have protective brown skin.

What is UV radiation?

Ultraviolet radiation is a member of the 'family' of electromagnetic radiations. Just as visible light can be spread out to form the spectrum from red to violet; electromagnetic radiation can be spread out to form the **electromagnetic spectrum**. UV appears just beyond the violet end of the visible spectrum.

The electromagnetic spectrum is a family of electromagnetic radiations that all travel at the same very high speed, 300 000 km/s, through space.

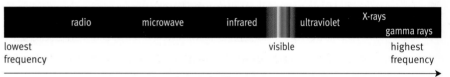

increasing frequency

We can think of electromagnetic radiation travelling as waves. Radio waves are the waves with the lowest frequencies. Gamma rays and X-rays have the highest frequencies.

Sunlight and skin

We need sunlight to fall on our skin. This is because human skin absorbs sunlight to make vitamin D. This nutrient strengthens bones and muscles. It also boosts the immune system, which protects you from infections. Recent research suggests that vitamin D can also prevent the growth and spread of cancers in the breast, colon, ovary, and other organs.

Darker skin makes it harder for the body to make vitamin D. So in regions of the world that are not so sunny there is an advantage in having fair skin. People with dark skin can keep healthy in less sunny countries if they get enough vitamin D from their food.

Balancing risks and benefits

People like sunshine. It can alter your mood chemically and reduce the risk of depression.

But is sunlight good for you? There is no simple answer. Over a lifetime, the risk of developing one type of skin cancer, malignant melanoma, is

Fair skin is good at making vitamin D. But it gives less protection against UV radiation. Melanoma is the worst kind of skin cancer. One severe sunburn in childhood doubles the risk of melanoma in later life.

1 in 91 (UK males) or 1 in 77 (UK females). These figures were calculated in 2009, but the incidence of skin cancer is increasing rapidly. However, if you try and avoid skin cancer by staying indoors, there are risks too.

Protecting your health involves reducing risks, whenever possible, and balancing risks against benefits.

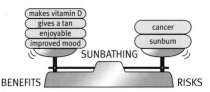

Many people sunbathe. They reckon the benefits outweigh the risks.

Skin cancer warnings ignored

Too much exposure to the Sun is dangerous. A Cancer Research UK survey found a worrying gap between how much people know about skin cancer and how little they actually do to protect themselves in the Sun.

Among 16–24-year-olds, 75% believed that exposure to the Sun might cause skin cancer. But only a quarter of this age group apply high-factor sunscreen as protection. And only a third cover up or seek shade from the Sun.

Key words

- ✔ **electromagnetic spectrum**
- ✔ **ultraviolet radiation (UV)**
- ✔ **factor**
- ✔ **outcome**
- ✔ **correlation**
- ✔ **cause**

Correlation or cause?

A study of 2600 people found that people who were exposed to high levels of sunlight were up to four times more likely to develop a cataract (clouding of the eye lens). Exposure to sunlight is possibly a **factor** in causing cataracts. Eye cataracts are an **outcome**. There is a **correlation** between exposure to sunlight and eye cataracts. But doctors do not say that exposure to sunlight will produce cataracts – it may not be the **cause**. There are other risk factors involved, such as age and diet.

Questions

1 Look at the diagram of the electromagnetic spectrum on the opposite page.
   a Which type of radiation has the lowest frequency?
   b Which colour of visible light has the highest frequency?

2 A person with dark skin moves to live in a region where there are few sunny days. Why should they try to spend a lot of time out of doors?

3 Exposure to sunlight increases your risk of developing skin cancer. List some benefits of staying indoors and avoiding direct sunlight. Suggest some risks of staying indoors.

4 Look at the information about the risk of melanoma. Who is more at risk of developing skin cancer? Suggest a reason why.

Find out about

- ✔ **sources of light, and the paths light follows**
- ✔ **how the ozone layer protects life on Earth**

Key words

- ✔ **source**
- ✔ **emits**
- ✔ **transmit**
- ✔ **reflect**
- ✔ **absorb**
- ✔ **ozone layer**
- ✔ **atmosphere**
- ✔ **CFCs**

Coloured materials added to glass can absorb some colours of light and transmit others.

A beautiful world

All radiation has a **source** that **emits** it. Then it has a journey. It spreads out, or 'radiates'. Radiation never stands still. Some radiation, at the end of its journey, causes chemical changes at the back of your eye. That radiation is visible light.

Some materials, like air, are good at **transmitting** light. They are clear, or transparent. On the way from the source to your eyes light can be **reflected** by other materials. The objects around you would be invisible if they did not reflect light.

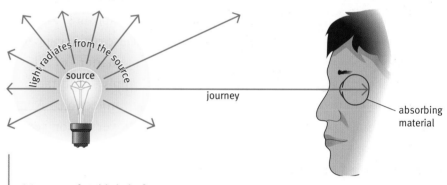

A journey of visible light from source to eye.

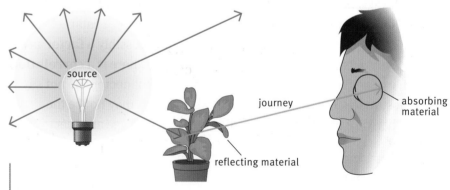

A journey of visible light, from source to reflector to eye.

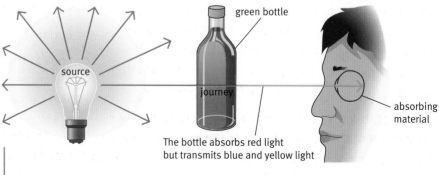

A journey from source to detector, but with absorption of light on the way.

An absorbing atmosphere

Radiation from the Sun contains infrared radiation, visible light, and ultraviolet radiation. Some of this radiation is transmitted through the atmosphere so that it reaches the ground. Fortunately for us, most of the harmful UV radiation is **absorbed** as the Sun's radiation passes downwards through the atmosphere. Life on Earth depends on the ozone layer absorbing UV radiation.

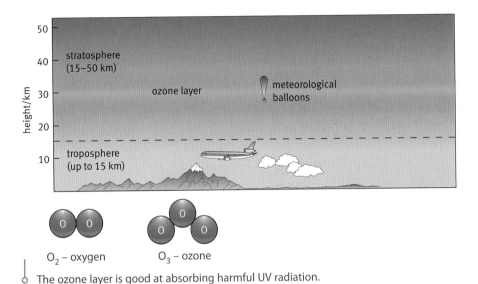

O_2 – oxygen O_3 – ozone

The ozone layer is good at absorbing harmful UV radiation.

The **atmosphere** is a mixture of gases, including oxygen. In the upper atmosphere some of the oxygen atoms combine in threes to make ozone. It makes an **ozone layer**.

The ozone layer is good at absorbing UV radiation. When UV radiation is absorbed its energy can break ozone and oxygen molecules, making free atoms of oxygen.

These chemical changes are reversible. Free atoms of oxygen in the ozone layer are constantly combining with oxygen molecules to make new ozone.

Ozone holes

Humans have created a problem. Some synthetic (man-made) chemicals, such as **CFCs** (chlorofluorocarbons) used in fridges, have been escaping into the atmosphere. They turn ozone back into ordinary oxygen. So more UV radiation reaches the Earth's surface. This happens strongly over the North and South Poles. More UV radiation can reach the Earth's surface through the 'hole in the ozone layer'.

The international community is now dealing with the problem. Aerosol cans no longer use CFCs. Old fridges have the CFCs carefully removed at the end of their working life. However, the ozone layer may take decades to return to its original thickness.

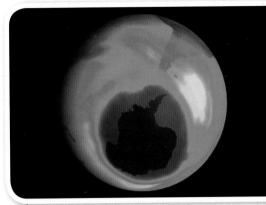

This image has been made by sensing ozone. Dark colours represent less dense ozone. There seems to be a 'hole' in the protective layer.

Old fridges waiting to have CFCs removed.

Questions

1 Can glass reflect light? Explain how you know.

2 Materials can transmit, reflect or absorb light. Which one of these is glass best at? How can this property of glass be changed?

3 Use the words – *source, reflect,* and *detector* to explain how you can see to read at night, using a torch.

4 Why is it important for life on Earth that the atmosphere absorbs most of the UV radiation that comes from the Sun?

When materials absorb electromagnetic radiation they gain energy. Exactly what happens depends on the type of radiation.

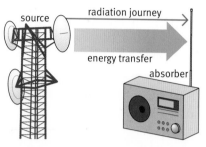

Metal aerials can absorb radio and microwave radiation. The process creates electrical patterns inside the metal.

Radiation can cause a varying electric current in a metal wire

Patterns of microwave and radio radiation can make patterns of electric current in radio aerials.

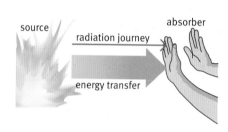

A fire transfers energy to the world around it. Surfaces in its surroundings, including people, absorb the radiation and gain the energy.

Radiation can have a heating effect

Radiation absorbed by a material may increase the vibration of its particles (atoms and molecules). The material gets warmer.

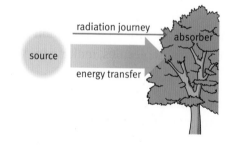

A leaf takes energy from the Sun's radiation so that photosynthesis can happen.

Radiation can cause chemical changes

If the radiation carries enough energy, the molecules that absorb it become more likely to react chemically. This is what happens, for example, in photosynthesis, and in the retinas of your eyes.

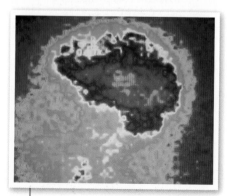
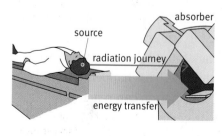

This medical image was made by a gamma camera. Each dot on the image was made by a single ionisation event.

Ionisation can damage living cells

If the radiation carries a large amount of energy, it can remove an electron from an atom or molecule, creating charged particles called **ions**. This process is called **ionisation**. The ions can then take part in other chemical reactions. Ionisation can damage living cells.

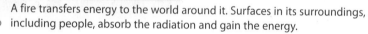

Radiation arrives in energy packets

Radiation is 'grainy'. It carries energy in small packets called **photons**:

- Sources emit energy photon by photon.
- Absorbers gain the energy photon by photon.

The energy deposited by a beam of electromagnetic radiation depends on both:

- the number of photons arriving at the absorber
- and the energy that each photon delivers.

Ionising and non-ionising radiation

Look back at the electromagnetic spectrum on page 166. Radiations with the highest frequencies have the photons with the highest energies. **X-ray** photons and **gamma ray** photons carry most energy. Radio photons carry least energy.

Sources of gamma rays, X-rays, and high-energy UV radiation pack a lot of energy into each photon. A single photon has enough energy to ionise an atom or molecule, by knocking off an electron. Gamma rays, X-rays, and some UV are **ionising radiations**.

Visible, infrared, microwave and radio radiations are all **non-ionising radiations** because a single photon does not have enough energy to ionise an atom or molecule. The main effect of these radiations is warming. The lower the photon energy is, the smaller the heating effect of each photon.

Lying in the sunshine, infrared, and visible radiations have a warming effect; UV radiation can initiate a chemical change that could start skin cancer.

Key words

- ✓ **ions**
- ✓ **ionisation**
- ✓ **photons**
- ✓ **ionising radiation**
- ✓ **non-ionising radiation**
- ✓ **X-rays**
- ✓ **gamma rays**

Question

1 There is radio radiation passing through your body right now.
   a Where does the radio radiation come from?
   b Why does it not have any ionising effect?
   c Does it have a heating effect? Explain.

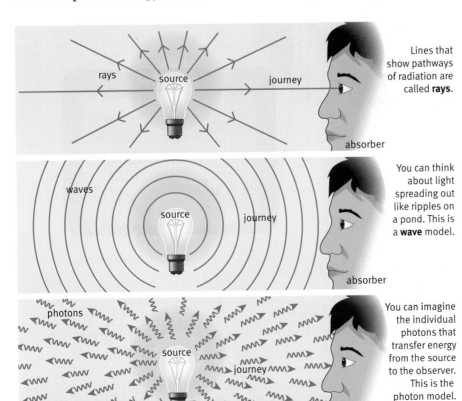

rays
source
journey
absorber

Lines that show pathways of radiation are called **rays**.

waves
source
journey
absorber

You can think about light spreading out like ripples on a pond. This is a **wave** model.

photons
source
journey
absorber

You can imagine the individual photons that transfer energy from the source to the observer. This is the photon model.

Radiation transfers energy. There are different ways of thinking about how it travels between source and absorber.

Find out about

- ✔ **reducing the risk from ionising radiation**
- ✔ **how ionising radiation can affect body cells**

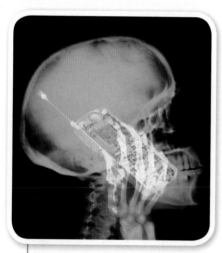

Both the health benefits and the risks of X-rays are well known. Using a mobile phone has benefits but uncertain risks.

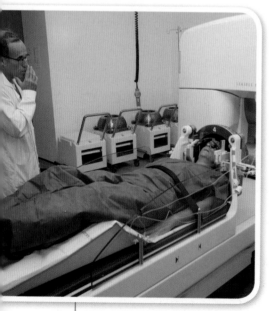

This patient is undergoing gamma radiotherapy; gamma radiation from the machine above him is directed towards a cancerous tumour in his body.

Using X-rays

X-rays were discovered in the 1890s. They soon caught on as a medical tool, and they have saved many lives. How do they work?

X-rays are produced electrically, in an X-ray tube. A beam of X-rays is shone through the patient and detected on the other side using an X-ray camera.

As the beam passes through the patient, it is partly absorbed. Bone is a stronger absorber than flesh, and so bones show up as 'shadows' on the final image. We say that the **intensity** of the beam is reduced as it is absorbed.

The intensity of a beam of electromagnetic radiation is the energy that arrives at a square metre of surface each second. This tells us how 'strong' the beam is.

The intensity of any radiation is less if you are further away from the source. This is because the radiation spreads out over a wider and wider area.

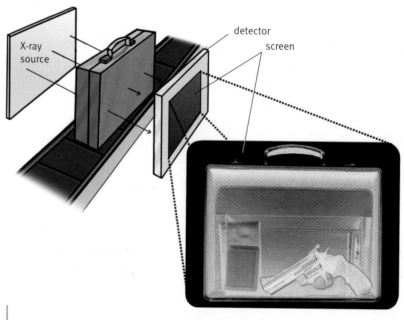

X-ray machines are used in airport security checks. Like bone, metal objects absorb X-rays strongly and so reduce the intensity of the beam.

Using gamma rays

Gamma radiation is similar to X-rays but it comes from **radioactive materials**. These are substances that emit radiation all the time – you can't switch them off.

Gamma radiation is used, like X-rays, for imaging a patient's internal organs. It is also used to destroy cancer cells.

Discovery of a correlation

Alice Stewart (see photo) and George Kneale carried out a survey on a large number of women and their children. They discovered a correlation between X-ray exposure of mothers during pregnancy and cancers in their children.

There is a plausible **mechanism** that could explain this correlation. X-ray photons can ionise molecules in your body, and is particularly risky if DNA molecules are affected. This can disrupt the chemistry of body cells, and cause cancer. So the link is more than just a correlation. X-rays can, in a few cases, cause cancer.

This study made doctors more cautious about using X-rays. The radiation is more damaging to cells that are rapidly dividing, such as a growing fetus or a small child. So the risks associated with X-rays for small children and pregnant women usually outweigh any benefit.

Reducing the risk

When a patient has an X-ray, the equipment and procedures are designed to keep the X-ray exposure to the minimum that still produces a good image. People who work with ionising radiation must also be protected from its effects.

There are several ways to reduce exposure to ionising radiation:

- **time:** the shorter the time of exposure, the less radiation is absorbed so the smaller the chance of damage to cells
- **distance:** intensity decreases as radiation spreads out from the source
- **shielding:** use materials such as lead and concrete, which absorb radiation strongly
- **sensitivity:** use a detector that is more sensitive so that less radiation is needed to produce an image.

UV photons have enough energy to change atoms and molecules, which can initiate chemical reactions in body cells. This can cause skin cancer. This is why it is advisable to cover up with clothes and sunscreen to reduce the risk on a sunny day.

Obituaries

Alice Stewart

Alice Stewart was a British doctor. She collected and analysed information from women whose children had died of cancer between 1953 and 1955. Soon the answer was clear. On average, one medical X-ray for a pregnant woman was enough to double the risk of early cancer for her child.

Key words
- ✓ **intensity**
- ✓ **radioactive materials**
- ✓ **mechanism**

Questions

1 a Name the three types of electromagnetic radiation that are also ionising radiations.
   b Which of these has the least energetic photons?

2 In the article about Alice Stewart, what is the **outcome** she is studying? What is the **factor** that might be causing this outcome?

3 a Why is the link between X-ray exposure during pregnancy and childhood cancer believed to be a 'cause' and not just a 'correlation'?
   b Why do doctors still use X-rays, despite this link?

Microwave ovens

A microwave oven uses **microwave radiation** to transfer energy to absorbing materials. Once the materials have absorbed the energy of the radiation, it ceases to exist as radiation. Molecules of water, fat, and sugar are good absorbers of microwave radiation. When they absorb microwave radiation, they vibrate strongly; in other words, their temperature rises.

A potato contains water, so it absorbs microwave radiation. The intensity of the radiation decreases as it passes into the potato. The particles of glass or crockery absorb very little energy from the radiation. It does not increase their vibrations, so bowls and mugs are not heated directly. They are heated by the hot food or drink inside them.

Absorption of energy by a potato. The radiation is gradually absorbed as it passes into the potato. It may not reach the middle of a very large potato. The plate absorbs very little energy from the microwave radiation.

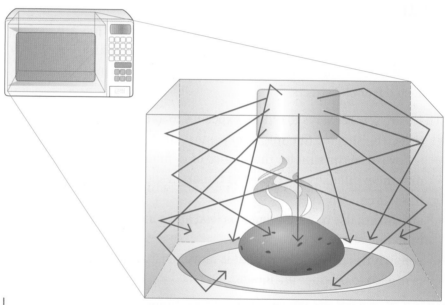

Transmit, absorb, reflect … Inside a microwave oven, materials like glass and pottery only partially transmit the radiation. The metal walls reflect it. Some substances, such as the water in a potato, absorb it.

Safety features

People contain water and fat, two absorbers of microwave radiation. So microwaves could cook you. The oven door has a metal grid to reflect the radiation back inside the oven. A hidden switch prevents the oven from operating with its door open.

How well cooked?

Any material that absorbs non-ionising radiation (such as microwaves) gets hot. The heating effect depends on the **intensity** of the radiation and its **duration** (the exposure time).

You control the amount of cooking in a microwave oven by adjusting:

- the power setting (to control the intensity)
- the cooking time (to control the duration of exposure).

Cooked brain?

Mobile phones use microwave radiation to send signals back and forth to the nearby phone mast. When you make a call, the fairly thick bone s of your skull absorb some of this radiation. But some reaches your brain and warms it, ever so slightly. Vigorous exercise has a greater heating effect.

There is no evidence that this exposure is harmful. However, some people use a hands-free kit so that the phone is further from their head. The radiation is less intense because it spreads outwards from the phone.

We're all radiators

Hot objects glow brightly. They emit visible light. In fact, cool objects emit electromagnetic radiation too. This is invisible **infrared** radiation. Even objects whose temperature is far below 0°C emit weak infrared radiation. This has a very low frequency.

The hotter an object is, the higher the frequency of the radiation it emits. Very hot objects, such as the hottest stars, emit radiation whose **principal** or main frequency is in the ultraviolet region of the electromagnetic spectrum.

A mobile phone stops radiating when you stop speaking. It also sends a weaker signal when you are close to the phone mast. That's to save the battery, but it also means that less radiation penetrates your head.

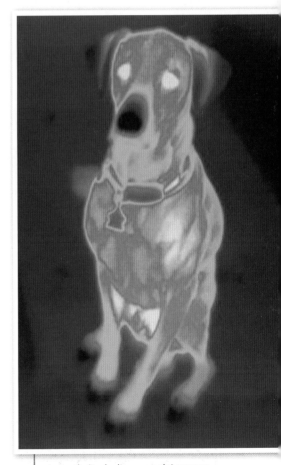

Animals (including people) are too cool to emit visible light, but a special camera can detect the infrared radiation they give out.

Questions

1 What radiations are on either side of microwave radiation in the electromagnetic spectrum?

2 Why doesn't microwave radiation cause ionisation?

3 Why is it important that the walls and door of a microwave oven reflect the microwave radiation?

4 Many people imagine that it is dangerous to live close to a mobile phone mast. Why might your exposure to microwave radiation from your phone be reduced if you lived close to a mast?

5 The damage caused by radiation depends on the energy of the photons. Energy of photons of electromagnetic radiation is *proportional to* the frequency of the radiation. Explain what is meant by 'proportional to.'

Find out about

- ✓ how the Earth is warmed by the Sun
- ✓ how the atmosphere keeps the Earth warm
- ✓ why the amount of CO_2 in the atmosphere is changing

Data collected by the meteorological Office shows that Bognor Regis gets more hours of sunshine than any other town in England. This data is more reliable than any individual's recollections.

Key words

- ✓ climate
- ✓ greenhouse effect
- ✓ greenhouse gases

Are summers now hotter and winters milder than they once were? This is a question about **climate**, or average weather in a region over many years. You cannot answer it from personal experience, because you can only be in one place at a time. And memory can be unreliable. Instead, you need to collect and analyse lots of data.

A comfortable temperature for life

The Earth's average temperature is about 15 °C, which is very comfortable for life. Why does it have this temperature?

Firstly, we are in orbit around the Sun. The Earth's surface absorbs radiation from the Sun, and this warms the Earth. At the same time, the Earth emits radiation back into space.

- During the day, our part of the Earth is facing the Sun. The Sun's radiation is absorbed by the Earth. It warms us up and the temperature rises.
- At night, we are facing away from the Sun. Energy radiates away into space and the Earth gets colder.
- The Earth is cooler than the Sun, so the radiation it emits has a lower principal frequency. It can be absorbed by the atmosphere.

Without the Sun to keep topping up the Earth's store of energy, the Earth's temperature would soon fall to the temperature of deep space, about −270 °C.

The greenhouse effect

Without its atmosphere, the Earth's average surface temperature would be –18 °C. That's how cold it is on the Moon. This warming of the Earth by its atmosphere is called the **greenhouse effect**.

Life on Earth depends on the greenhouse effect. Without it, the Earth's water would be frozen. Water in its liquid form is essential to life.

Greenhouse gases

If the atmosphere consisted entirely of the commonest gases (nitrogen and oxygen), there would be no greenhouse effect.

Questions

1 Personal experience does not provide reliable evidence of climate change. Why not?

2 Which of the following gases found in the Earth's atmosphere are *not* greenhouse gases?: nitrogen, methane, oxygen, carbon dioxide, water vapour?

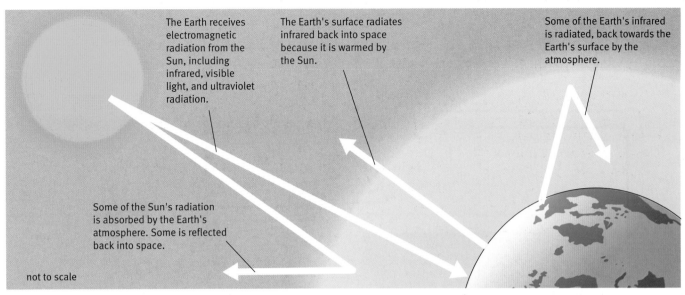

The Earth receives electromagnetic radiation from the Sun, including infrared, visible light, and ultraviolet radiation.

The Earth's surface radiates infrared back into space because it is warmed by the Sun.

Some of the Earth's infrared is radiated, back towards the Earth's surface by the atmosphere.

Some of the Sun's radiation is absorbed by the Earth's atmosphere. Some is reflected back into space.

not to scale

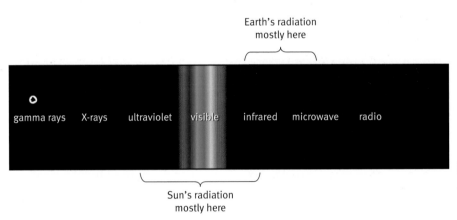

Earth's radiation mostly here

gamma rays   X-rays   ultraviolet   visible   infrared   microwave   radio

Sun's radiation mostly here

There is an energy balance between radiation coming in and going out of the atmosphere. The atmosphere lets in infrared radiation from the Sun, but prevents the infrared emitted by the Earth from escaping. This is because the Earth's radiation has lower frequencies than the Sun's, and these are absorbed by the atmosphere.

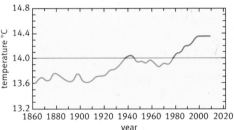

The Earth's surface temperature has risen over the past 150 years (data from weather stations).

Tiny amounts of a few other gases make all the difference. Carbon dioxide, methane and water vapour absorb some of the Earth's infrared radiation. They are called **greenhouse gases**.

Past temperatures

Weather stations have kept temperature records for over a century. The graph opposite shows the results.

There is a clear pattern. The Earth's average temperature has been rising since 1800. This conclusion is supported by evidence from Nature's own records (growth rings in trees, ocean sediments, air trapped in ancient ice).

Most climate scientists think that carbon dioxide (CO_2) in the atmosphere is causing the rise in temperatures. Why?
* Temperatures and CO_2 levels have risen at the same time.
* Evidence from the distant past suggests that temperature and CO_2 levels go up and down together.
* Scientists can explain how CO_2 in the atmosphere absorbs radiation and raises the temperature.

Questions

3 All of the statements about CO_2 and the Earth's average temperature describe correlations. Which statement is also about cause and effect?

4 Look at the temperature graph. Use the graph to describe how the Earth's surface temperature has changed over the past 150 years.

The carbon cycle

Carbon dioxide is a greenhouse gas that plays a key role in global warming. Industrial societies produce CO_2 as never before.

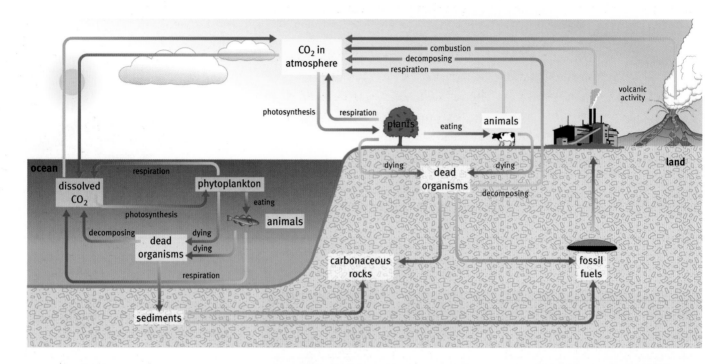

The Earth's crust, oceans, atmosphere, and living organisms all contain carbon. Carbon atoms are used over and over again in natural processes. The **carbon cycle** describes the stores of carbon and processes that move carbon.

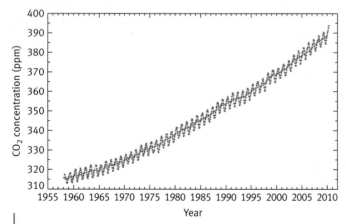

Carbon dioxide concentrations have been recorded at Mauna Loa in Hawaii since 1958. They rise and fall each year, but the overall trend has been an increase of about 1.5 ppm per year since 1980.

Carbon dioxide in the atmosphere

Hundreds of millions of years ago, the amount of CO_2 in the atmosphere was much higher than it is today. Green plants made use of CO_2 and released oxygen. This made life possible for animals. Eventually, lots of carbon was locked up underground in the form of fossil fuels, as well as carbonaceous rocks such as limestone and chalk.

For thousands of years CO_2 levels were stable. Plants absorbed CO_2 during photosynthesis, then animals and decomposers returned it to the atmosphere during respiration. Humans burned wood, but that was balanced by new trees growing. Two hundred years ago, the industrial revolution changed all that as fossil fuels were burned in increasing quantities.

Human activities release carbon

People want to live comfortably. In some parts of the world, many feel they have a right to processed foods, unlimited clean water and electricity, manufactured goods, and bigger houses and cars. All of these things require energy.

But whenever fossil fuels (coal, oil, and gas) are burned, they increase the amount of carbon dioxide in the atmosphere. Methane, another greenhouse gas, is produced by grazing animals and from rice paddies. Cutting down or burning forests (**deforestation**) also releases CO_2, and reduces the amount removed by photosynthesis.

Although methane is the more effective greenhouse gas, carbon dioxide produced by human activities has a bigger effect. This is because the amount of CO_2 is so huge – thousands of millions of tonnes each year. This is why there are international agreements on reducing 'carbon emissions'.

Motor vehicles are a major source of greenhouse gas emissions.

This power station supplies enough electricity for a major city. Every day it uses several trainloads of coal and sends thousands of tonnes of carbon dioxide into the atmosphere.

Air transport is a big user of fossil fuels. Aviation fuel is cheap because it is untaxed, unlike petrol for cars.

People in the UK use more energy for keeping buildings warm than for anything else.

Questions

5 Study the diagram of the carbon cycle opposite.
   a List six processes that release CO_2 into the atmosphere.
   b List two processes that remove CO_2 from the atmosphere.
   c Which of the above processes has changed so that the amount of CO_2 in the atmosphere is increasing?

6 Look at the graph of CO_2 levels opposite. Explain its shape: why does it go up and down every year? Why is the long-term trend upwards?

7 Forest land can be cleared for farming by burning trees. Explain why tree burning increases the amount of carbon dioxide in the atmosphere.

Key words
✓ **carbon cycle**
✓ **deforestation**

Find out about

- ✓ possible effects of climate change
- ✓ ways of reducing the amount of CO_2 released into the atmosphere

Nature's records

The polar ice caps are frozen records of the past. In parts of the Antarctic, ice made from annual layers of snow is four kilometres thick. That ice contains tiny bubbles of air, a record of the atmosphere over 800 000 years. It shows that climate has always changed. There have been ice ages and warm periods.

But temperatures have never increased so fast as during the past 50 years.

Natural factors change climates

Over the long term, natural factors cause climate change. For example:
- the Earth's orbit changes the distance to the Sun by small amounts
- the amount of radiation from the Sun changes in cycles
- volcanic eruptions increase atmospheric CO_2 levels.

These factors cause much slower changes than we are seeing today, but they still must be taken into account when scientists try to determine whether human activities are causing the climate to change.

Climate modelling

The atmosphere and oceans control climates. Climate scientists use **computer models** to predict the effects of increasing CO_2 levels.

What is a computer model? Climate models are similar to the models used for day-to-day weather forecasting. They use everything climate scientists know about how the atmosphere and oceans behave. For example, we know that, at present, the Earth absorbs about 1% more energy from the Sun than it radiates back into space. This extra energy warms the oceans, increasing the rate at which water evaporates, and the amount of water vapour in the atmosphere. The computer models can calculate how this will affect temperatures around the world.

Computer models are tested using data about the Earth's climate in the past. If they can correctly account for this data, it is more likely that their predictions for the future will be accurate.

However, the further we try to look into the future, the greater the uncertainty in our predictions.

Alarming predictions

What these models show is alarming.
- Human activities now contribute more to climate change than natural factors.

The tiny bubbles in this slice of ancient ice from Antarctica contain air trapped hundreds of thousands of years ago.

An 'ice core' like this stores a record of the Earth's changing atmosphere over hundreds of thousands of years.

Key word
- ✓ computer models

- Future emissions of greenhouse gases are likely to raise global temperatures by between 2 and 6 °C during your lifetime.
- If CO_2 concentration rises much further, climate change may become irreversible.
- To stabilise climates, carbon emissions would need to be reduced by at least 70% globally.
- In the UK, winters will become wetter and summers drier.

Climate change is a slow process. It may take 20 to 30 years for climates to react to the extra CO_2 already in the atmosphere. So global temperatures are guaranteed to rise by 2 °C. Ice will continue to melt, and sea levels continue to rise, for the next 300 years or so, even if humans today stopped producing any CO_2 at all.

Effects of climate change

Human societies depend on stable climates. The risks associated with global warming are enormous.

Extreme weather: There are likely to be more extreme weather events (violent storms, heat-waves). This is because higher temperatures will cause greater convection in the atmosphere, and more evaporation of water from both oceans and the land.

Rising sea levels: Water in the oceans will expand as it gets hotter, so sea level will rise. In addition, continental ice sheets, such as in Antarctica and Greenland, may melt, adding to the volume of the oceans. Low-lying land will be flooded, causing a particular problem for people who live in river deltas such as the Ganges delta in Bangladesh, or on low-lying islands.

Drought and desertification: Reduced rainfall may make it impossible to grow staple crops in some areas. Tropical areas may become drier, leading to the expansion of deserts such as the Sahara.

Health: Malaria will spread if mosquitoes can breed in more places.

Climate-change sceptics

Thousands of climate scientists have contributed to our understanding of how human activities are affecting the climate. They publish their results in scientific journals and test each other's ideas.

Some scientists and many other non-specialists have challenged aspects of this work. These people are sometimes called 'climate sceptics' or 'deniers'. Despite these challenges new evidence usually supports climate scientists' ideas.

These maps, based on satellite photographs, show how the area of the Arctic ice sheet decreased between 1980 and 2007.

Questions

1 Sea levels are rising. Give two reasons why they are likely to continue rising in the future.

2 Explain why rising temperatures will give rise to more violent storms.

3 Which **two** effects of global warming may make it difficult to grow some food crops in particular regions?

4 Make a list of the scientific uncertainties mentioned on these pages.

Paying the price

The world's poorest countries will be least able to deal with the effects of climate change. Their people are the most vulnerable. Yet it is the wealthier countries that are responsible for the problem.

The bar chart shows this:

- The height of each block in this graph shows how much CO_2 is emitted per person each year – North Americans emit most.
- The area of each block shows the total emissions for each region – people in Europe and North America are responsible for over 60% of emissions.

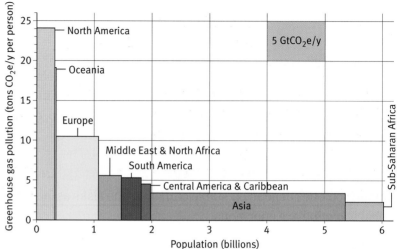

The UK is at risk too

The UK climate is kept mild by the Gulf Stream, a warm current from the Caribbean that flows towards Europe across the North Atlantic. There is evidence that this current slowed down in the past, making the UK an icy place. There are signs that the Gulf Stream may be slowing again.

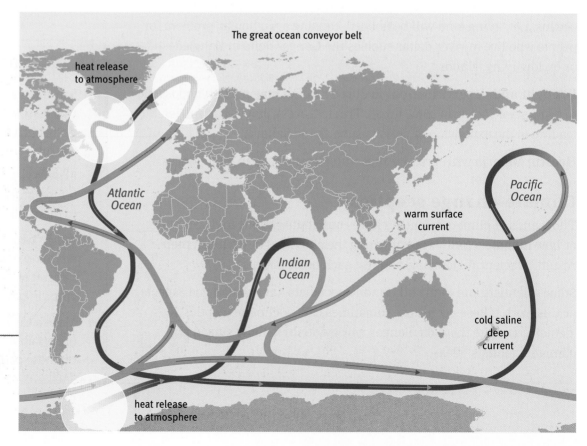

The great ocean conveyor belt

The Earth has a giant 'conveyor belt' system of ocean currents. It helps to warm land in northern latitudes.

What can governments do?

The UK government aims to reduce greenhouse gas emissions:

- 20% by the year 2020
- 80% by 2050.

The baseline is 592 million tonnes emitted in 1990. The graph shows there has been a small decrease in emissions.

It can be difficult for governments to change people's behaviour so that they produce less carbon dioxide. Democratic governments are sensitive to public opinion because they face election every few years. They may find it difficult to do what's best for the long term. People may protest and businesses may fight to protect their profits.

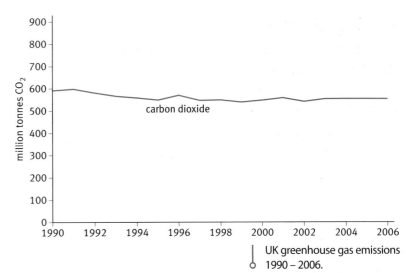

UK greenhouse gas emissions 1990 – 2006.

Technological solutions

There are many different proposals for using technology to reduce the amount of CO_2 entering the atmosphere. However, these may take a long time to develop and put into operation – and their outcomes may be uncertain.

Science to the rescue

Several solutions have been proposed that might get around the difficulties of reducing carbon emissions.

- Spread iron granules on the southern oceans. This would help the growth of plankton, which take dissolved CO_2 from the ocean. The oceans would remove more CO_2 from the atmosphere.
- Capture the CO_2 produced at power stations. Then compress it into a liquid and pump it into disused oil reservoirs beneath the sea-bed.
- Cement production counts for 5% of the greenhouse gases produced in Europe and America and more than 10% in China. A new type of 'eco-cement' absorbs CO_2 while setting and goes on absorbing CO_2 for years afterwards.

These are currently being tried and evaluated.

Extract from a popular science magazine.

Questions

5 Look at the four sources of emissions described in Section F . For each one, suggest what the government could do to reduce carbon emissions.

6 Look at the bar chart on the opposite page. What does the wide and low block for Asia show?

7 Do you think people should rely on technical solutions, like those suggested in the science magazine? Justify your answer.

Find out about

- ✔ **using electromagnetic waves to transmit information**
- ✔ **analogue and digital signals**

We use aerials for sending and receiving radio signals to mobile phones. They are high up so that there is a good 'line of sight' between phone and aerial.

Key words

- ✔ **information**
- ✔ **carrier wave**
- ✔ **analogue signal**
- ✔ **digital signal**

What is 'information'?

A mobile phone can store images (photographs), sounds (music and voice messages), text (messages), and numbers. Images, sounds, text, and numbers are all forms of **information**.

The phone can receive and transmit information, because it is part of a telephone network. It can also store and process information, for example, the user can amend a text message or play a game.

The phone is not full of pictures, sounds, and text. It stores the information electronically. This makes it easy to process the information, display it on the screen, or send it to another user.

Information paths

Mobile phones use microwaves. These are radio waves with the highest frequencies (and shortest wavelengths). Other radio waves are used to broadcast radio and TV programmes.

Radio waves and microwaves can travel for long distances through the air because they are only weakly absorbed by the atmosphere. This means that microwaves are also suitable for communicating with spacecraft far out in space.

Information is also transmitted along optical fibres, using visible light, infrared, or ultraviolet radiation. The fibres are made of high-purity glass, so that the radiation can travel many kilometres without being significantly absorbed.

Optical fibres have made possible cable television systems and the high-speed phone lines that are used by the internet.

Carrying information

Information is sent from place to place using a **carrier wave**. The radio waves, visible light or infrared can form the carrier wave. It must be modified to include the information. A wave carrying information is called a signal:

carrier wave + information = signal

You can send information at night using a flashing torch. The light from the torch is the carrier wave; the on-off flashes are the coded information. This is very similar to what happens in an optical fibre. A series of on-off pulses of light travel through the fibre and are received at the other end. An on-off signal, like the one below, is a **digital signal**. Just two symbols are needed to represent the signal: 0 (off, no pulse) and 1 (on, pulse).

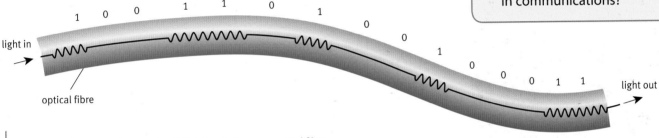

Pulses of light are used to send a digital signal along an optical fibre.

Coding information

All types of information can be converted into digital signals. A sound wave is an example. Speak into a microphone and your sound waves are converted into a varying voltage. The graph shows how the voltage might change during a fraction of a second.

The electrical signal has the same pattern as the original sound wave. A continuous variation like this is an **analogue signal**.

The table under the graph shows how this changing voltage can be converted into a digital signal. The voltage is sampled many times per second and the height of the graph (in volts) is coded as a binary number. These are the 0s and 1s that are used to switch the carrier wave on and off.

Images (such as photographs) can be coded by dividing them up into tiny dots (pixels). Each pixel is given a numerical value, which codes its colour and brightness.

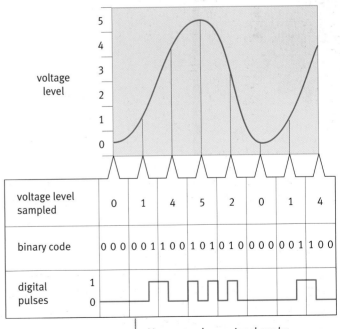

voltage level sampled	0	1	4	5	2	0	1	4
binary code	0 0 0	0 0 1	1 0 0	1 0 1	0 1 0	0 0 0	0 0 1	1 0 0

How an analogue signal can be converted into digital pulses. (Real systems use more levels and a much faster sampling rate.)

aerial — speaker

processor/memory

keypad

microphone

The main components of a mobile phone.

Each pixel in the digital image has a number that gives information about the colour of that part of the picture. The bigger the choice of colours, the greater the range of numbers needed.

Radiation carries information

Are you receiving me?

If you send a message to a friend by flashing a torch, they must be able to decode it. Similarly, a mobile phone must be able to decode the digital signals it receives.

Inside the phone is a microprocessor that can do this. It simply reverses the process shown in the diagram on the previous page. It takes the binary codes and converts them into a varying voltage. This is sent to the earphone and you hear the sound.

More information

The amount of information needed to store an image or sound is measured in bytes (B). It takes about 1 megabyte (1 MB) to store one minute's worth of music. The memory of a hand-held mp3 player can vstore a hundred gigabytes of information. That is several weeks' worth of music, or thousands of photographs.

Each pixel in a digital image has information about the colour of that part of the picture. In a black-and-white image the pixel has a number to describe the brightness of the dot – from white through shades of grey to black. In a colour image the information also has to say what colour the dot is – this may be one of 256 colours or more than a million possible colours and shades. The colour image contains more information, which takes more time to send and more memory to store.

The pictures show two versions of the same picture.

Each pixel in the black-and-white image has one of 256 shades of grey between white and black; the information can be stored in 271 kB of memory. In the colour image each pixel can take over a million different colours and needs 74 MB of memory.

Mobile phones contain precious metals including gold, which can be recovered during recycling. Many people upgrade regularly as the technology improves.

The more information that is stored about an image or sound, the better quality it is.

Advantages of digital transmission

For transmitting information such as sounds and pictures, digital signals have several advantages over analogue ones.

- Digital signals can be processed by microprocessors (as in computers and phones).
- Digital information can be stored in memories that take up little space (as in computers, phones, and mp3 players).
- Digital signals can carry more information every second than analogue ones.
- Digital signals can be delivered with no loss of quality. Analogue signals lose quality, which cannot be restored.

Here is the reason for the last point. All signals get weaker as they travel along. **Noise** ('interference') also gets added in. A noisy analogue signal carrying music might sound blurry, scratchy, and distorted. But with digital signals, these effects can be corrected.

The diagram shows how. The noisy digital signal is passed through a regenerator. This is an electronic circuit that removes the noise. It can do this because it knows that the value of the signal must be either 0 or 1. A value close to 0 is corrected to 0, a value close to 1 is corrected to 1. It would take a very large amount of noise before it became difficult to tell the 0s and 1s apart!

Key words
- ✓ byte (B)
- ✓ noise

Questions

4 Which contains more information, a 100 kB image file or a 1 MB sound file?

5 Explain why a 10 MB image file will produce a better picture than a 1 MB file of the same image.

6 Why is it possible to remove noise from digital signals but not from analogue ones?

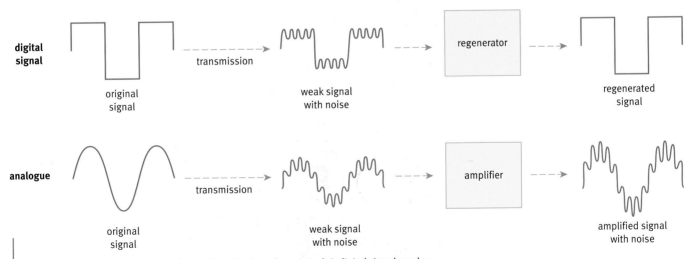

After transmission, a signal is weaker and noisier than the original. A digital signal can be 'cleaned up' by a regenerator. If an analogue signal is amplified to return it to its original height, the noise gets amplified as well.

Mobile phones are useful and they are fun. But are they dangerous?

A mobile phone sends out signals that are detected by a nearby mast. This is how the phone connects to the network.

Because the aerial is close to the user's ear, the user's head will absorb some of the energy of the microwave radiation. The amount of energy absorbed depends on the intensity of the radiation and the length of time of the exposure.

Absorbed radiation has a heating effect so, as we saw on page 175, the user's head is slightly heated by the radiation it absorbs. This depends on the number of photons absorbed and the energy each delivers.

Radiation spreading out

The intensity of the radiation coming from a mobile phone is greatest as it leaves the phone. Because the radiation spreads out in all directions, it rapidly gets weaker the further away it travels from the phone.

Similarly, the radiation from the mast is most intense as it leaves the mast. It is more intense than the radiation from a phone. By the time it reaches a distant phone, it is very weak.

People have concerns about radiation from phone masts. If you stood right next to the mast, the heating effect of the radiation absorbed by your body could be quite noticeable.

Fortunately, you cannot get that close. Phone masts are designed so that their radiation is shaped like the beam of light from a lighthouse. If you stand directly under a mast, its radiation is much weaker than the radiation from your phone.

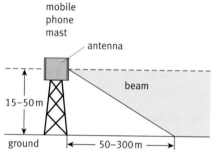

The microwave beam from a mobile phone mast.

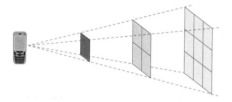

Twice as far away from the phone, the area is four times as great, so only a quarter of the radiation falls on each small square.

Some people think that road travel is less risky than going by train. But almost 3000 people die each year in UK roads. Far fewer people die on the railways. This is an example of the difference between a **perceived** risk and an **actual** risk.

People may not accurately estimate a particular risk. People tend to overestimate the risks of things with invisible effects, like radiation. Also, they overestimate the risks of things where they feel less in control. For example, people worry more about flying than cycling, although many more people are killed or injured in cycling accidents.

Health studies

Over 50 million people in the UK use mobile phones. Few worry about any unknown risks. People like the benefits a mobile phone brings. Research has so far failed to show that there are any harmful effects.

To look for any harmful effects, scientists compare a sample of mobile phone users with a sample of non-users. Does one group show a higher rate of cancer, for example?

Can we have confidence in the results?

The news often has reports of studies that compare samples from two groups, to see if a particular factor or treatment makes a difference. To judge whether **we can have confidence in the results** of studies like this, there are two things worth checking.

What to check ...	... and why
Look at how the two samples were selected. Can you really be sure that any differences in outcomes are really due to the factor claimed?	A study to find whether mobile phone use caused cancer would need to compare samples of users and non-users. The samples should match as many other factors as possible, for example, similar numbers of young people in each sample. This is because the development of brain tumours might be age-related. Samples should be selected randomly, so that other factors, such as genetic variability, are similar in both groups.
Were the numbers in each sample large enough to give confidence in the results?	With small samples, the results can be more easily affected by chance. Larger sample sizes give a truer picture of the whole population. The effect of chance is more likely to average itself out.

How great is the risk?

Health outcomes are often reported as relative risks. For example, 'people exposed to high levels of sunlight were four times as likely to develop eye cataracts'.

- If your risk was one in a million, it rises to four in a million – not a worry!
- If your risk was 5 in 100, it rises to 20 in 100 – worth avoiding!

And some people might be more at risk than others, for reasons of family history or lifestyle.

Key words
- ✓ reliable
- ✓ perceived risk
- ✓ actual risk

Questions

1   Is there a health risk from low-intensity microwave radiation from mobile phone handsets and masts? (The answer is neither 'Yes' nor 'No'.)

2   a   Look at the first row of the table. What outcome and what factor that might cause it are being studied?

   b   Describe a second way that the samples should be matched in this study.

   c   Explain why it is important to match the two samples.

Science Explanations

In this module you will learn about the electromagnetic spectrum, different types and sources of radiation, and how it can be both useful and dangerous.

You should know:

- how to think about any form of radiation in terms of its source, its journey path, and what happens when it is absorbed, transmitted, or reflected
- that a beam of electromagnetic radiation delivers energy in 'packets' called photons
- the parts of the electromagnetic spectrum, in order of their photon energies
- what different parts of the electromagnetic spectrum can be used for
- two factors that affect the energy deposited by a beam of electromagnetic radiation
- how the intensity of an electromagnetic beam changes with distance
- that the three parts of the electromagnetic spectrum with highest photon energies are ionising
- why ionising radiation is hazardous to living things
- how people can be protected from ionising radiation
- how microwaves heat materials, including living cells
- about the features of microwave ovens that protect users
- that sunlight provides the energy for photosynthesis and warms the Earth's surface
- how photosynthesis affects which molecules are in the atmosphere
- what the greenhouse effect is (and be able to identify greenhouse gases)
- how to use the carbon cycle to explain how green plants and decomposers have kept the carbon dioxide concentration constant for thousands of years
- how the atmosphere's ozone layer protects living organisms from ultraviolet radiation
- about global warming and its possible effects on agriculture, weather, and sea levels
- that computer models provide evidence that human activity causes global warming
- that radio waves and microwaves carry information for radio and TV through the atmosphere and through space
- that infrared and visible light waves carry information along optical fibres
- that a sound wave can be superimposed onto an electromagnetic wave, and that this signal can be carried, and then decoded to produce a copy of the original sound
- that this coding can be an analogue signal, which varies continuously, or a digital signal, which is a series of pulses
- the advantages of digital signals in reducing noise, ease of storage, and manipulation of the stored signals.

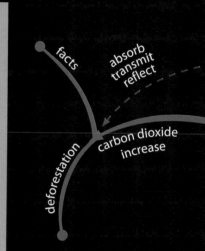

facts

absorb
transmit
reflect

carbon dioxide
increase

deforestation

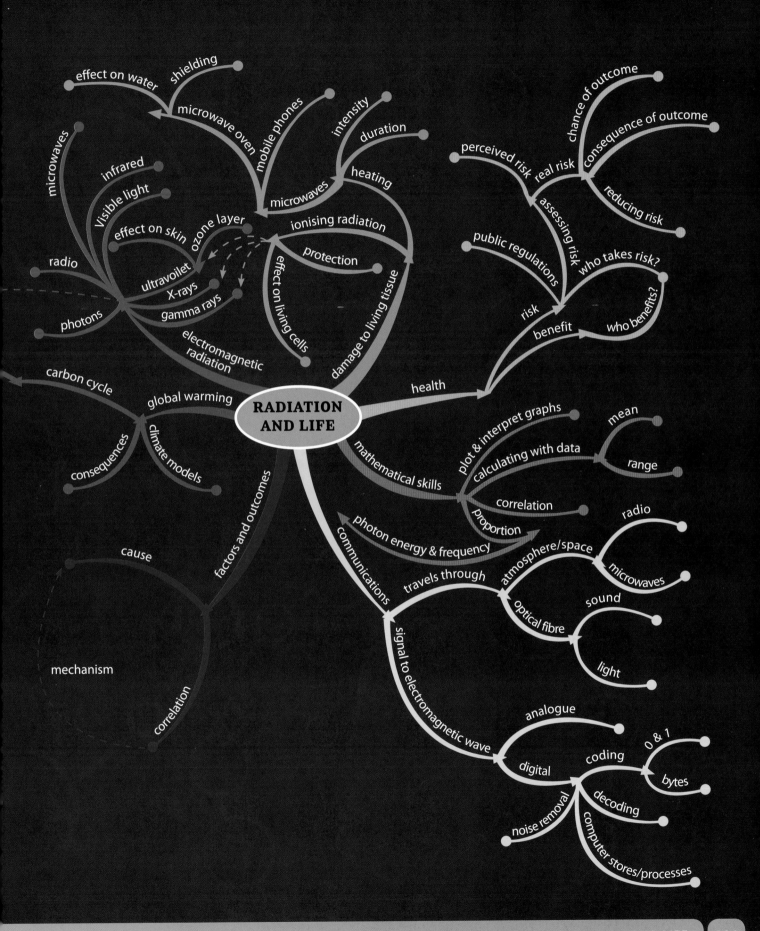

Ideas about Science

Besides developing an understanding of the electromagnetic spectrum, it is important to recognise the difference between correlation and cause, and to assess the risks and benefits associated with the electromagnetic spectrum.

Factors and outcomes may be linked in different ways, and it is important to distinguish between a correlation, where a change in one is linked to a change in the other, from a cause, where there is a method by which the factor is responsible for the outcome.

In the context of the electromagnetic spectrum, you should be able to:
- suggest and explain everyday examples of correlation
- identify a correlation from data, from a graph, or from a description
- suggest factors that might increase the chance of a particular outcome, but not invariably lead to it
- identify that where there is a mechanism to explain a correlation, scientists are likely to accept that the factor causes the outcome.

Everything we do carries some risk, and new technologies often introduce new risks. It is important to assess the chance of a particular outcome happening, and the consequences if it did, because people often perceive a risk as being different from the actual risk: sometimes less, and sometimes more. A particular situation that introduces risk will often also introduce benefits, which must be weighed up against that risk.

You should be able to:
- identify risks arising from scientific or technological advances
- interpret and assess risk presented in different ways
- discuss risk, taking into account both the chance of it occurring and the consequence if it did

- discuss both the risks and benefits of a course of action, taking into account of who takes the risk and who benefits
- distinguish between real risk and perceived risk
- suggest why people are willing (or reluctant) to take certain risks
- discuss how risk should be regulated by governments and other public bodies.

Review Questions

1 **a** Copy out and complete the diagram of the electromagnetic spectrum. Use words from this list.

microwaves　　　　**gamma rays**　　　**ultraviolet**

radio waves		infrared	visible light		X-rays	

b Where in the electromagnetic spectrum do the photons have the most energy?

c Which **one** of the following is **not** ionising radiation?

　　gamma rays　　　**microwaves**　　　**ultraviolet**　　　**X-rays**

2 The graph shows how the percentage of carbon dioxide in the atmosphere has changed over the past 300 years.

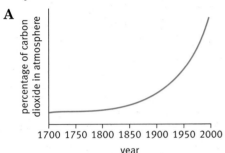

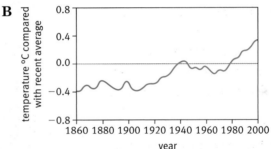

a Describe the trend in graph A and explain the scientific reasons for what the graph shows.

b Graph B shows how the average temperature of the atmosphere has changed over the past 140 years. Use the two graphs to explain the meaning of **correlation**.

3 Match the wave diagrams to the statements opposite. Diagrams may be used once, more than once, or not at all.

A 　　　D

B 　　　E

C 　　　F

a A sound wave is an analogue wave.

b The sound wave is converted into a digital code. The digital signal is sent as a series of short pulses.

c Digital signals can be transmitted with higher quality than analogue signals. As the signal is transmitted, it decreases in intensity and picks up noise.

d When the signal is received it is amplified.

e The signal is cleaned up to remove the noise.

f The digital signal is then decoded to reproduce the original sound wave.

B3 Life on Earth

Why study life on Earth?

There are over 30 million species of living things on Earth today. Where do they all come from? Why is there so much variety? Is that variety important? Can we learn to look after life on Earth better for future generations? These are the big questions we ask science to answer. Scientists think life began on Earth 3500 million years ago. The first simple organisms have developed and changed, and many species have become extinct.

What you already know

- Characteristics are determined by genetic information passed on from both parents.

- Individuals with the same genetic information may vary.

- Some characteristics are influenced by environmental factors.

- Photosynthesis is the source of biomass in plants.

- Plants need mineral salts for growth.

- The distribution of organisms in an environment is affected by environmental factors.

- Organisms only survive in a habitat where they have all the essentials for life and reproduction.

- Organisms show adaptations to environmental conditions.

- Food webs show feeding relationships.

Find out about

- how different species depend on each other

- how life on Earth is evolving

- how scientists developed an explanation for evolution

- why some species become extinct, and whether it matters.

The Science

Fossils and DNA provide evidence for how life on Earth has evolved. Simple organisms gradually change to form new species.

All life forms depend on their environment and on other species for their survival. Ultimately all life depends on the Sun's energy, and on nutrients, which are recycled through the environment.

Ideas about Science

Today most scientists agree that evolution happens. But nobody thought this 200 years ago. Developing new explanations takes evidence and imagination. Even then, the explanation will change as we collect new evidence and test the ideas. The variety of species on Earth is a valuable resource, which humans depend on. Scientists can help us devise ways to use natural resources in a more sustainable way.

You can usually see the differences between different kinds of living things on Earth. But there are also a lot of similarities, even between living things that don't look the same. For example, almost all living things use DNA to pass on information from one generation to the next.

Classification – working out where we belong

Scientists use the similarities and differences between living things to put them into groups. You've probably come across this idea before. It's called **classification**. The biggest group that humans belong to is the kingdom *Animalia* (the Animal Kingdom). The smallest is *Homo sapiens*, or human beings. *Homo sapiens* is our **species** name.

Classification names are in Latin, so that everyone can use the same name for something. It doesn't matter what languages two people speak, they can always use the same Latin name.

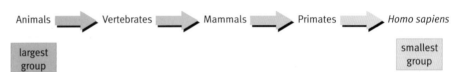

Animals ➡ Vertebrates ➡ Mammals ➡ Primates ➡ *Homo sapiens*

largest group — smallest group

You are most closely related to other members of *Homo sapiens*. But you belong to these other groups as well. Vertebrates are all the animals that have skeletons with backbones. All mammals have five-digit limbs. Primates have five-fingered hands and feet, shoulder joints that can move in all directions, and forward-facing eyes in bone sockets. Down the classification, the groups contain fewer organisms that have more and more characteristics in common.

What makes a species?

Scientists define a species as a group of organisms so similar that:

- they can breed together
- their offspring can also breed (they are **fertile**).

Human skin cells and cells in these butterfly wings use the same chemical reaction to make pigment.

horse

donkey

mule

Horses and donkeys do look very similar. But their offspring are infertile. So horses and donkeys are different species.

Horses and donkeys are good examples to explain species. They can breed together and produce offspring called mules. But mules are **infertile**. Horses and donkeys look pretty similar, but they are different species.

All members of a species are not the same. You only need to look around your classroom to see that this is true. This variation is caused by a mixture of genes and environment. It is very important in evolution. You'll find out more about this later.

The art of survival

All of the 30 million or so species presently alive on Earth are successful survivors. They have features that help them survive in their environments. These features are called **adaptations**.

Cactus plants live in hot, dry, desert conditions. They survive because they are adapted to this environment. They store water inside their stems and have large root systems to get water from deep in the soil. Many cactus plants have hard spines instead of leaves.

Fish have special adaptations that enable them to live in water. A fish absorbs oxygen dissolved in water. The oxygen diffuses from the water into the fish's blood across the large surface area of the gills. A streamlined body and a smooth surface helps the fish move through the water with little resistance. Sets of fins keep the fish balanced in the water. A swim bladder gives the fish buoyancy in the water.

Cacti and fish are both adapted to the habitat where they live.

Questions

1 What species do you belong to?

2 Explain why horses and donkeys are different species.

3 Give a reason for each of the cactus's adaptations.

Key words
- species
- fertile
- infertile
- adaptations

Find out about

- ✔ **why some species become extinct**
- ✔ **how organisms are interdependent**

Over the past few million years, many species of plants and animals have lived on Earth. Most of these species have died out. When all the members of a species die out it is **extinct**.

There is fossil evidence of at least five mass extinctions on Earth. Now we seem to be at the beginning of another.

Around the world over 12 000 species of plants and animals are at risk of extinction. They are endangered.

Where an animal or plant lives is called its **habitat**. Any quick changes in their habitat can put them at risk of extinction.

Changes in the environment

All living things need factors like water and the right temperature to survive. Rising temperatures are changing many habitats. This global warming is putting many species at risk.

New species

New species moving into the habitat can put another species at risk of extinction.

- Animals and plants compete with each other for the things they need. Two different species that need exactly the same things cannot live together. One wins the **competition** for resources like food and shelter.
- The new species could be a **predator** of the species already living there.
- If the new species causes **disease**, it could wipe out the native population.

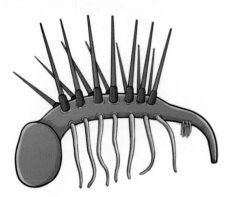

Fossil from burgess shale. This amazing-looking species lived 505 million years ago. It is extinct but scientists think it may be the ancestor of crabs and centipedes.

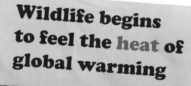

Wildlife begins to feel the heat of global warming

Six regions were studied, representing 20% of the Earth's land area.

A large international study says that up to a quarter of the species on Earth face extinction from global warming.

Red squirrels used to live all over the UK. Now the larger American grey squirrels have taken over most of their habitats.

In the 1960s, the virus that causes Dutch Elm disease came to the UK. It destroyed most of the UK elm population.

Going hungry

Plants and animals need other species in their habitat. For example, in this food chain spiders eat caterpillars.

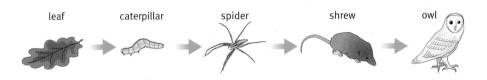

leaf · caterpillar · spider · shrew · owl

So if the caterpillars all died, the spiders could be at risk. That could also endanger the shrew and the owl.

The food web

Most animals eat more than one thing. Many different food chains contain the same animals. They can be joined together into a **food web**. This shows the **interdependence** of living organisms.

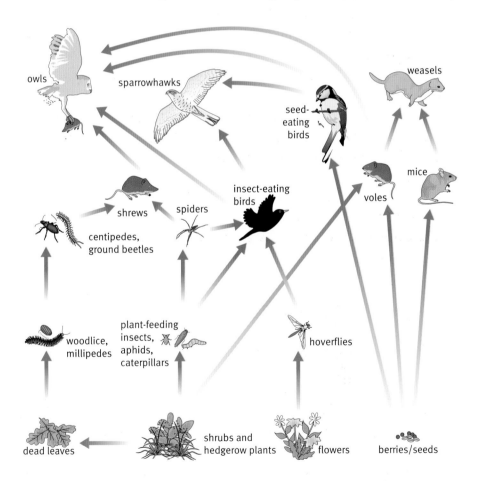

owls · sparrowhawks · weasels · seed-eating birds · shrews · spiders · insect-eating birds · voles · mice · centipedes, ground beetles · woodlice, millipedes · plant-feeding insects, aphids, caterpillars · hoverflies · dead leaves · shrubs and hedgerow plants · flowers · berries/seeds

A new animal coming into a food web can affect plants, animals, and microorganisms already living there. The loss of one organism will also affect other organisms in the food web.

Questions

1 Look at the food web on this page.
   a Name two different animals competing for the same food source.
   b A disease kills all the flowering plants. Explain what happens to the number of hoverflies.
   c Mink move into the habitat. They eat voles.
      i The number of mice decreases. Explain why.
      ii Explain what would happen to the number of caterpillars.

2 Explain what is meant by extinct.

3 Name two things that:
   a plant species may compete for
   b animal species may compete for.

Find out about

- ✔ **how organisms depend on the Sun's energy**
- ✔ **how energy and nutrients pass through food webs**

Nearly all organisms are ultimately dependent on energy from the Sun. The Sun provides energy to keep the Earth's atmosphere warm and drive the production of food chemicals.

Food chains like the one on page 199 all follow the same pattern. They start with plants, the **producers**.

Plants capture energy from sunlight. They use the energy to build organic compounds such as glucose from carbon dioxide and water. This is the process of **photosynthesis**.

Energy from sunlight is stored in these new compounds, which make up the plants' cells. The compounds can be broken down in **respiration** to release energy. The energy stored is passed to other organisms as the plants are eaten or decompose.

Plants trap only about 1–3% of the light energy that reaches their leaves in new plant material. This might sound small, but remember that the Sun's energy output is enormous. So this 1–3% is still enough energy to power life on Earth.

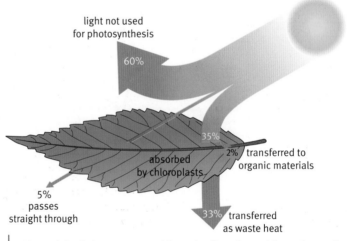

light not used for photosynthesis

60%

35%

2% transferred to organic materials

absorbed by chloroplasts

5% passes straight through

33% transferred as waste heat

Most of the light energy reaching a leaf is reflected from the surface, is transferred as waste heat, or passes straight through the leaf. Only a small percentage is absorbed by the chloroplasts where photosynthesis takes place.

Energy transfer

Animals have to eat. They cannot make their own food so they need to take in organic molecules. They are **consumers**, and they break down food molecules in respiration.

Some of the energy released by respiration is used for growth, where the food molecules become part of the structure of new cells. Animals also use energy released by respiration for other life processes, for example, keeping warm.

The sunbather enjoys the warmth of the Sun, but does not rely directly on the Sun's energy for his life processes. The plant is harnessing light energy to drive food production.

On average only about 10% of the energy at each stage of a food chain gets passed on to the next level. The rest:

- is used for life processes in the organism, such as movement
- escapes into the environment as heat energy
- is excreted as waste and passes to decomposers
- cannot be eaten and passes to decomposers

This means that the number of organisms usually gets smaller at each level of an ecosystem. The food chain is limited to only a few levels. Usually there are only the producers, primary consumers, secondary consumers, and tertiary consumers.

The diagram opposite shows the efficiency of energy flow through an ecosystem.

Energy also flows into **decomposers** and **detritivores** as they feed on dead organisms and waste material in the ecosystem. Bacteria and fungi are decomposers. Detritivores are small animals like woodlice.

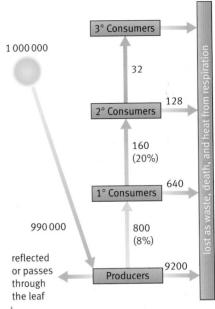

1 000 000

3° Consumers

32

2° Consumers

128

160 (20%)

1° Consumers

640

990 000

800 (8%)

9200

reflected or passes through the leaf

Producers

lost as waste, death, and heat from respiration

Energy flow through an ecosystem.

Woodlice, earthworms, millipedes, and insect larvae (maggots) are examples of detritivores.

Bacteria and fungi are decomposers. This photograph shows fungus growing on a dead weevil.

Key words

- ✓ producers
- ✓ photosynthesis
- ✓ consumers
- ✓ respiration
- ✓ decomposers
- ✓ detritivores

Questions

1 Look at the food web on page 199. Identify two:
   a producers
   b primary consumers
   c secondary consumers
   d tertiary consumers.

2 Explain why the energy in a producer will not all be transferred to the primary consumer in the same food chain.

3 Calculate the percentage efficiency of energy transfer between the secondary and tertiary consumers in the ecosystem above (shown in the diagram at the top of the page).

Find out about

- ✔ **how carbon and nitrogen are recycled through the environment**
- ✔ **how environmental change can be measured using living and non-living indicators**

Key words

- ✔ **combustion**
- ✔ **carbon cycle**
- ✔ **phytoplankton**
- ✔ **photosynthesis**
- ✔ **respiration**
- ✔ **decomposition**

Questions

1 List ways in which carbon dioxide is:
   a added to the atmosphere
   b taken out of the atmosphere.

2 How is human activity causing the amount of carbon dioxide in the atmosphere to rise?

3 Before 1970 the species of phytoplankton called *Ceratium trichoceros* was only found off the south coast of England. It is now found in the waters off Scotland. Suggest what environmental change is likely to have occurred to affect where this phytoplankton lives.

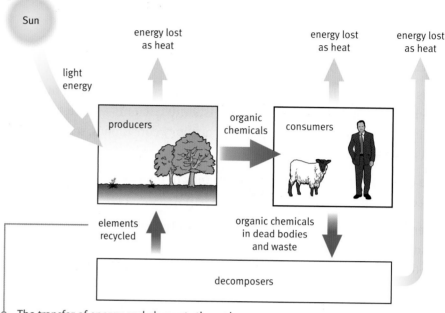

The transfer of energy and elements through an ecosystem.

Energy stored in the chemicals making up cells is passed on to other organisms along food chains. In this way energy moves through the ecosystem. This is similar to how elements like carbon and nitrogen move through the ecosystem. But there is a big difference.

Carbon and nitrogen are always being recycled in an ecosystem.

Recycling carbon

There is only a certain amount of carbon on Earth. Much of the carbon is in molecules that make up the bodies of living things. A lot is also in the atmosphere and oceans as carbon dioxide, and in molecules of fossil fuels.

Carbon dioxide is taken out of the atmosphere by **photosynthesis**. The carbon is used to produce glucose molecules. The glucose is broken down during **respiration**. This releases carbon dioxide back into the atmosphere.

When an animal or plant dies its organic compounds are broken down by microorganisms; this is called **decomposition**. The carbon and nitrogen atoms become part of a new organism in the system.

Carbon dioxide is also added to the atmosphere by the **combustion** (burning) of wood and fossil fuels, such as oil, gas, coal, and petrol. The stages in recycling carbon on Earth are shown in the diagram of the **carbon cycle** on the opposite page.

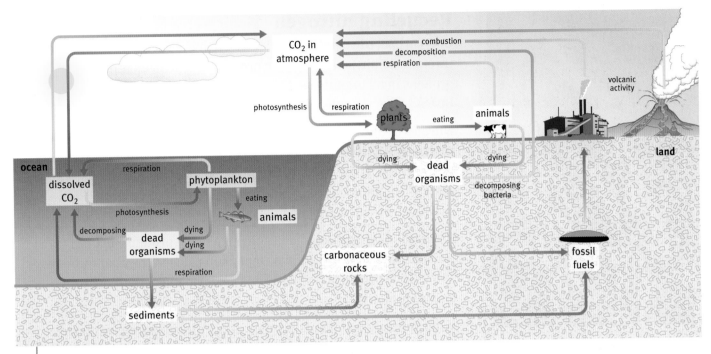

The carbon cycle.

Measuring environmental change

If the amount of carbon dioxide released into the air does not balance the amount taken up by photosynthesis, atmospheric carbon dioxide levels change.

Most scientists agree that the average level of carbon dioxide in the atmosphere is rising. The official figure in 2010 was 0.04% carbon dioxide. It is expected to be 0.05% by the end of the century. The current value is about 40% higher than any value measured over the past 800 000 years.

It is thought that the rise in carbon dioxide levels is linked to rises in the global temperature of the Earth. This is called global warming.

This climate change can be measured by looking at the impact on living organisms. **Phytoplankton** are tiny floating plants found in seawater. Their numbers and patterns of distribution are affected by the impact of climate change on water temperature, ocean mixing, and nutrients levels in the water.

Many species are threatened with extinction if conditions change beyond their ability to adapt.

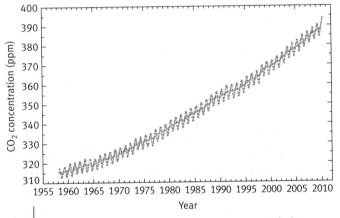

Carbon dioxide concentrations have been recorded at Mauna Loa in Hawaii since 1958. They rise and fall each year, but the overall trend has been an increase of about 1.5 ppm per year since 1980.

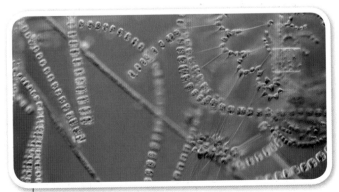

The distribution and abundance of species of phytoplankton (small drifting aquatic plants) change as the temperature of surface water rises.

Recycling nitrogen

Like carbon, nitrogen is recycled through the environment, passing through the air, soil, and living organisms. Nitrogen is part of the proteins in all organisms. Plants cannot take up nitrogen from the air. Instead, they take up nitrogen compounds, such as nitrates, from the soil.

Nitrates in the soil are made using nitrogen from the air. This is called **nitrogen fixation**. **Nitrogen-fixing bacteria** can turn nitrogen gas into nitrates. These bacteria live in the soil and inside swellings on the roots of leguminous plants, such as peas, beans, and clover. The swellings are called root nodules.

Plants take up nitrates and use them to make proteins. Primary consumers (herbivore animals) eat plants, digest the plant proteins, and use the products (called amino acids) to make their own animal proteins.

When animals and plants die they decay. **Decomposer bacteria** break down the proteins in dead organisms. This releases nitrates back into the soil so they are available for plants to absorb once again. Nitrogen in waste from excretion (urine and faeces) is also recycled in this way.

Nitrogen-fixing bacteria live in root nodules on these bean roots.

The nitrogen cycle.

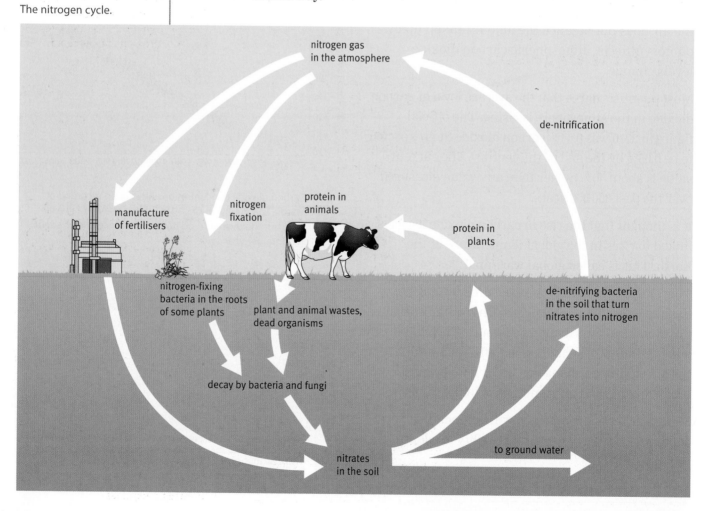

nitrogen gas in the atmosphere

de-nitrification

manufacture of fertilisers

nitrogen fixation

protein in animals

protein in plants

nitrogen-fixing bacteria in the roots of some plants

plant and animal wastes, dead organisms

de-nitrifying bacteria in the soil that turn nitrates into nitrogen

decay by bacteria and fungi

nitrates in the soil

to ground water

Denitrifying bacteria break down nitrates in the soil and release nitrogen gas back into the air. This process is known as **denitrification**. All the steps that make up the **nitrogen cycle** are shown in the diagram opposite.

Measuring environmental change

Farmers often add nitrogen to their fields in the form of chemical fertiliser (often ammonium nitrate). This helps make the soil more fertile and increases plant growth. But this can also have a negative impact, causing environmental change.

Chemical fertiliser is very soluble in water. Water rich in nitrate can drain from fields into streams, rivers, and lakes, raising nitrate levels in the water. High levels can cause the rapid growth of microscopic organisms (called plankton), form in massive 'algal blooms'.

Sometimes these blooms can cause sickness to people or animals drinking the water. Decomposer bacteria break down dead plankton. The bacteria respire using up oxygen dissolved in the water. Animals that need high levels of oxygen, such as **mayfly larvae**, cannot survive in water that contains high levels of nitrates.

Measuring nitrate levels and monitoring invertebrates allows changes in water quality to be measured.

Monitoring air quality

Nitrogen compounds can also pollute the air. Air quality can be monitored by studying the types of lichens surviving in a particular area.

Lichens are unusual organisms made up of a fungus growing with a green alga. The alga provides nutrients for the fungus through photosynthesis, while the fungus provides protection for the alga from drought and UV light from the Sun.

Some species of lichen can grow in areas with high levels of nitrogen compounds in the air, while others are very sensitive to nitrogen compounds and will only grow in places with no air pollution. Sensitive species are usually more feathery types. They become locally extinct when air pollution rises.

Key words

- ✓ nitrogen fixation
- ✓ nitrogen-fixing bacteria
- ✓ decomposer bacteria
- ✓ denitrifying bacteria
- ✓ denitrification
- ✓ nitrogen cycle
- ✓ mayfly larvae
- ✓ lichens

Monitoring invertebrates like these mayfly larvae, also known as mayfly nymphs, allow changes in water quality to be measured.

The golden shield lichen can live in areas with high levels of nitrogen, especially ammonia.

The heather rags lichen is very sensitive to nitrogen pollution in the air, so is rarely found close to roads in big cities.

Questions

4   Explain how nitrogen gets from the air into plants.

5   A gardener notices that the feathery-type lichens have disappeared from her garden. What change in the environment may have occurred?

6   Make a table summarising the different types of bacteria involved in the nitrogen cycle and stating the role of each.

Explaining similarities – the evidence for evolution

Scientists agree that life on Earth began about 3500 million years ago. Life started from a few simple living things. This explains why living things have so many similarities.

These simple living things changed over time to produce the enormous variety of different species of living things on Earth today. The changes also produced many species that are now extinct. This process of change is called **evolution**, and it is still happening today.

What evidence is there for evolution?

Fossils are the remains of dead bodies of living things preserved in rocks. They are very important as evidence for evolution. Almost all fossils found are of extinct species. This is more than 99% of all species that have ever lived on Earth.

How reliable is fossil evidence?

Conditions have to be just right for fossils to develop. Only a very few living things end up as fossils. So there are gaps in the fossil record.

Scientists have collected millions of fossils. This huge amount of evidence has helped to build up a picture of evolution. For example, scientists examined fossils of a dinosaur called *Sinosauropteryx* and discovered orange and white rings down its tail, made of primitive feathers. This provided evidence that the first birds developed from small meat-eating dinosaurs.

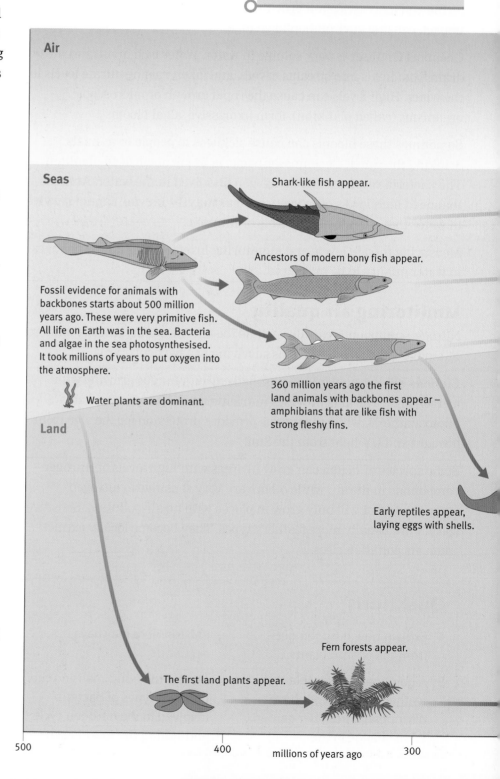

Air

Seas

Shark-like fish appear.

Ancestors of modern bony fish appear.

Fossil evidence for animals with backbones starts about 500 million years ago. These were very primitive fish. All life on Earth was in the sea. Bacteria and algae in the sea photosynthesised. It took millions of years to put oxygen into the atmosphere.

Water plants are dominant.

Land

360 million years ago the first land animals with backbones appear – amphibians that are like fish with strong fleshy fins.

Early reptiles appear, laying eggs with shells.

Fern forests appear.

The first land plants appear.

500   400   300

millions of years ago

What other evidence do we have for evolution?

Scientists also compare the DNA from different living things. The more similar the DNA of two living things, the more closely related they are. This helps scientists to classify them, work out where different species fit on the evolutionary tree, and make sense of the variety of species.

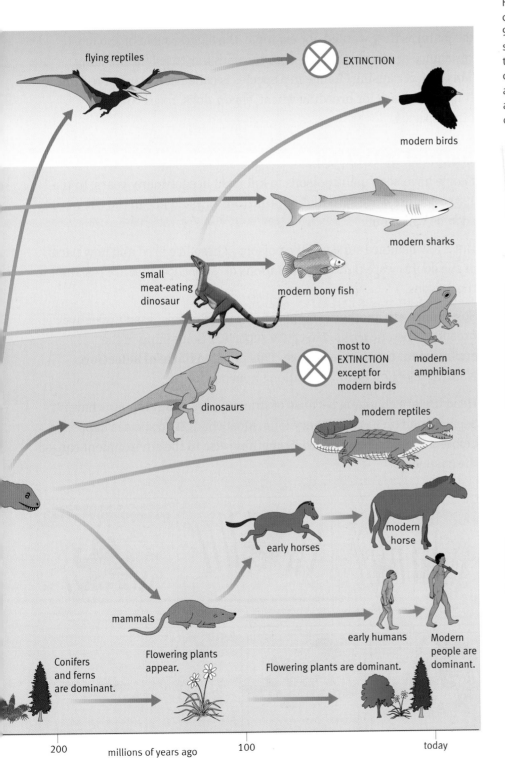

flying reptiles

EXTINCTION

modern birds

modern sharks

small meat-eating dinosaur

modern bony fish

most to EXTINCTION except for modern birds

modern amphibians

dinosaurs

modern reptiles

early horses

modern horse

mammals

early humans

Modern people are dominant.

Conifers and ferns are dominant.

Flowering plants appear.

Flowering plants are dominant.

200	100	today

millions of years ago

Humans and chimpanzees have over 98% of their DNA the same. They were thought to share a common ancestor about 6 million years ago. They are both classified as Primates.

Humans and mice are both mammals. About 85% of their DNA is the same. They shared a common ancestor about 75 million years ago.

Questions

1 What percentage of all life on Earth is alive now?

2 Name two types of evidence that scientists use as evidence for evolution.

3 Describe the changes that have occurred in the evolution of modern birds.

Find out about

- ✓ **how evolution happens – natural selection**
- ✓ **how humans have changed some species**

Selective breeding has produced tulips with different coloured flowers.

Head lice are quite common. They feed on blood.

Key words

- ✓ selective breeding
- ✓ populations
- ✓ natural selection

Evolution did not just happen in the past. Scientists can measure changes in species that are happening now. They expect evolution will continue in the future. Humans can alter the evolution of some species.

Selective breeding

Early farmers noticed that there were differences between individuals of the same species. They chose the crop plants or animals that had the features they wanted, for example, the biggest yield or the most resistance to diseases. These were the ones they used for breeding. This way of causing change in a species is called **selective breeding**. It has been used for breeding wheat, sheep, dogs, roses, and many other species.

Natural selection

People have been using poisons to kill head lice for many years. In the 1980s, doctors were sure that **populations** of head lice in the UK would soon be wiped out.

But a few headlice survived the poisons. These lice bred and now parts of the country are fighting populations of 'superlice' that are resistant to the poisons.

So headlice are another example of change. But this wasn't selective breeding – no one *wanted* to cause superlice. Something in the environment caused the change. This is called **natural selection**. Natural selection is how evolution happens.

Head lice are changing because of human beings. But humans haven't been around on Earth for very long. Most changes to species happened before human beings arrived. Something else in the environment caused the change.

For many years people used the same shampoo to kill head lice.

A few head lice in the population were able to survive. Their cells were probably able to break down the poison.

'Superlouse' was more likely to breed than the head lice killed by the poison.

Eggs laid by 'Superlouse' hatched into lice that also survived the poison.

These lice spread to other people and bred.

The number of resistant lice in the population increased. People couldn't get rid of their head lice.

Scientists developed a new poison to kill the head lice.

The cycle began again – and the species changed a little more.

Steps in natural selection

① *Living things in a species are not identical. They have variation.*

Ancestors of modern giraffes had variation in the length of their necks.

② *They compete for things like food, shelter, and a mate. But what if something in the environment changes?*

Food supply became scarce. The giraffes competed for food.

③ *Some will have features that help them to survive. They are more likely to breed. They pass their genes on to their offspring.*

Taller giraffes were able to eat more food, so were more likely to survive and breed. They passed on their features to the next generation.

④ *More of the next generation have the useful feature. If the environment stays the same, even more of the following generation will have the useful feature.*

Over many generations, more giraffes with longer necks were born. The number of taller giraffes in populations increase.

Treating HEAD LICE

Your Local Health Authority issues a directive, known as a rotational policy, every two to three years to inform people about which type of insecticide is currently recommended for use in your area.

The rotational policy is intended to prevent head lice becoming resistant to treatment – in other words, to help ensure that the treatments available continue to be effective in killing lice.

Questions

1   How does evolution happen?

2   Copy and complete the table below to compare selective breeding and natural selection.

Steps in selective breeding	Steps in natural selection
Living things in a species are not all the same.	Living things in a species are not all the same.
Humans choose the individuals with the feature that they want.	
These are the plants or animals that are allowed to breed.	
They pass their genes on to their offspring.	
More of the next generation will have the chosen feature.	
If people keep choosing the same feature, even more of the following generation will have it.	

3   Explain what is meant by a population.

4   Read the extract from a leaflet about head lice. Explain how this rotational policy stops the evolution of resistant populations of head lice.

5   Natural selection is sometimes described as 'survival of the fittest'. How good a description of natural selection do you think this is?

Find out about

- ✓ **how Darwin explained evolution**
- ✓ **how explanations get accepted**

Today most scientists agree that evolution happens. But evolution wasn't always as well accepted. A very important person in the story of evolution was Charles Darwin. His ideas were a breakthrough in persuading people that evolution happens.

Darwin's big idea

Darwin worked out how evolution could happen. He explained how natural selection could produce evolution. But he didn't come up with this idea overnight. It took many years.

Charles Darwin was born in 1809. He was interested in plants and animals from a young age. When he was 22, Darwin was given the chance to sail on HMS *Beagle*. The ship was on a five-year, round-the-world trip to make maps.

Journey of the *Beagle*

The *Beagle* stopped at lots of places along the way. At each stop Darwin looked at different types of animals and plants. He collected many specimens and made lots of observations about what he saw. He recorded this data in notes and pictures.

One place the *Beagle* stopped at was the Galápagos Islands, near South America. As he travelled between the different islands, Darwin noticed variation in the wildlife. One thing Darwin wrote about after his trip was the different species of finches living on the Galápagos Islands.

Darwin on HMS *Beagle*.

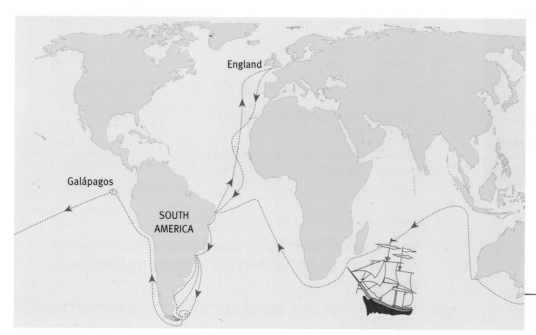

The *Beagle* stopped at different places around the world.

England

Galápagos

SOUTH AMERICA

The famous Galápagos finches

Each species of finch seemed to have a beak ideally suited for eating particular things. For example, one had a beak like a parrot for cracking nuts. Another had a very tiny beak for eating seeds. It was as if the beaks were adapted to eating the food on each different island.

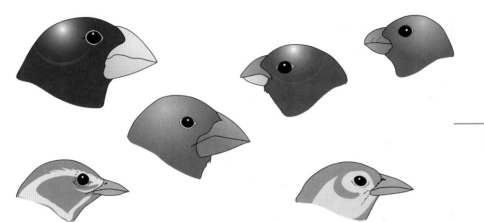

Different species of finch.

In his notes, Darwin started to ask himself a question. He wondered if all the different finches could have evolved from just one species.

What was special about Darwin?

Darwin wasn't the first scientist to think that evolution happens. His own grandfather was one of several people who had written about it earlier. But most people at the time didn't agree with evolution. Darwin was the first person to make a strong enough argument to change their minds.

He started by looking at lots of living things. He made many observations that he would use as evidence for his argument.

- He thought about the evidence in a way that no-one had done before. He was more creative and imaginative.
- He came up with an idea to explain *how* evolution could happen – natural selection.

Darwin showed his notes to a friend, Thomas Huxley. Huxley was also a scientist. When he read them, Huxley said: "How stupid of me not to have thought of this first!"

> One might really fancy that from an original paucity of birds in this archipelago one species had been taken and modified for different ends.

Charles Darwin, *The Voyage of the Beagle*, 1839.

> I look to the future to young and rising naturalists who will be able to view both sides of the question with impartiality.

Charles Darwin, *On the Origin of Species*, 1859.

Questions

1 Darwin made many observations about different species. How did he record his data?

2 What personal qualities did Darwin show that helped him develop his explanation of natural selection?

More evidence back home

Back in England, Darwin moved to a new home in Kent. For 20 years he worked on his idea of natural selection. He exchanged letters with other scientists in different parts of the world. All the time, Darwin was looking for more evidence to support his ideas.

His new home, Down House, had some pet pigeons. They had many different shapes and colours. But Darwin knew they all belonged to the same species. So he realised that:

- animals or plants from the same species are all different – there is **variation.**

Darwin found more evidence for natural selection at home.

Too many to survive

Next, Darwin realised that:

- there are always too many of any species to survive.

He came to this conclusion after reading the work of a famous economist, Thomas Malthus. At the end of the 18th century, Britain's population was growing very fast. Malthus pointed out that the numbers of any species had the potential to increase faster than any increase in their food supply. He predicted that the human population would grow too large for its food supply, and that poverty, starvation, and war would follow.

All the plants or animals of one species are in **competition** for food and space. A lot of them don't survive.

Darwin put these ideas together. He saw that some animals in a population were better suited to survive than others. They would breed and pass on their features to the next generation. This natural selection could make a species change over time. Darwin had explained how evolution could happen.

Elephants usually reproduce from age 30–90. Darwin worked out that after 750 years, if they all survived, there would be nearly 19 million elephants from just one pair!

Owing to this struggle for life, any variation, however slight, if it be in any way profitable to an individual of any species, will tend to the preservation of that individual, and will generally be inherited by its offspring. I have called this principle, by which each variation, if useful, is preserved, by the term of Natural Selection.

Charles Darwin, *On the Origin of Species*, 1859.

Key words
- ✔ variation
- ✔ competition

Same data, different explanations

Other scientists also saw that living things were different. They also saw fossils that showed changes in species. Fifty years before Darwin published his ideas, a French scientist called Lamarck had written a different explanation to Darwin's. He said that animals changed during their lifetime. Then they passed these changes on to their young. He used the example of a giraffe, as shown opposite.

Giraffe evolution explained by Lamarck.

Why was Darwin's explanation better?

A good explanation does two things:
- It accounts for all the observations.
- It explains a link between things that people hadn't thought of before.

Lamarck's explanation said that 'nature' had started with simple living things. At each generation, these got more complicated. If this kept happening, simple living things, like single-celled animals, should disappear. So his idea didn't account for some observations, for example, why simple living things still existed on Earth.

Darwin's idea could better account for these observations. It also linked together variation and competition, which hadn't been done before.

Lamarck's ideas may sound a bit daft now, but he was a good scientist trying to explain changes in species.

Why was Darwin worried about his explanation?

Darwin was worried about how people would react. He wrote his idea of natural selection into a book. Then he wrapped the manuscript in brown paper and stuffed it in a cupboard under the stairs. He wrote a note for his wife explaining how to publish the manuscript when he died. It stayed in the cupboard for almost 15 years.

Questions

3 How did Darwin try to get more evidence to support his ideas?

4 What two things make a good explanation?

5 What two things did Darwin link together to work out his explanation of natural selection?

I never saw a more striking coincidence. If Wallace had my manuscript sketch written out he could not have made a better abstract!

Charles Darwin, in a letter to the geologist Charles Lyell.

People agreed with Darwin's observations. But they didn't agree with his explanation.

At the 1860 British Association for the Advancement of Science (BA) meeting, Huxley and Hooker argued in favour of Darwin's theory.

On the Origin of Species

Then, in 1856, Darwin received a letter from another scientist, Alfred Russell Wallace. In it Wallace wrote about the idea of natural selection. Darwin was stunned. He gave Wallace credit for what he had done, and the two of them published a short report of some of their ideas. But now Darwin wanted to publish his full book before Wallace, or anyone else, beat him to it.

The now famous *On the Origin of Species* was published in November 1859. This book caused one of the biggest arguments in the history of science.

Many people in Victorian England disagreed with the idea of evolution of life by natural selection.

They were unhappy about the idea that humans were related to apes.

Why did people start to believe in evolution?

The British Association for the Advancement of Science meets every year. Scientists meet to share their ideas. In 1860, many scientists argued against Darwin's idea.

But two of his friends, Thomas Huxley and Joseph Hooker, defended it. They were very good scientists. They were also very good at speaking in public. So they helped to change many people's minds about natural selection.

The end of the story?

Natural selection was a good explanation. However, there were three big problems with it. But it wasn't Darwin's opponents who spotted these. It was Darwin himself.

Firstly, he knew that the record of fossils in the rocks was incomplete. At that time it was even more difficult to trace changes from one species to another than it is today. New fossil evidence has been found since then to support the idea of natural selection.

Secondly, the age of the Earth had not been worked out accurately enough. In Darwin's time it was thought to be about 6000 years old. So there didn't seem to have been enough time for evolution of complex organisms to have taken place. Scientists now have evidence from studying decay of radioactive atoms in rocks to show that the Earth is about 4.5 billion years old.

The last problem was in two parts:

- Darwin could not explain why all the living things in one species were not all the same. Where did variation come from?

- Also, he could not explain how living things passed features on from one generation to the next.

Both of these puzzles would have been easier for Darwin to answer if he had known about genes. Scientific discoveries since his time have allowed other scientists to do this.

At the same time that Darwin was writing *On the Origin of Species*, an Austrian monk called Gregor Mendel (1822–84), was breeding pea plants. From his experiments he discovered dominant and recessive alleles – different versions of the same genes. Mendel's work explained how features were passed on. He sent a copy of his work to Darwin, but Darwin didn't realise how important it was. Mendel's work was largely ignored until 16 years after his own death.

The debate goes on

In 1996, the late Pope John Paul II, head of the Roman Catholic Church, acknowledged Darwin's ideas with the words: "... new scientific knowledge leads us to recognise more in the theory of evolution than hypothesis."

People continue to debate evolution. Because many of them have strong personal beliefs that are affected by this idea, the debate is unlikely to stop anytime soon.

Questions

6 Most people in the 1800s disagreed with natural selection. What evidence did they have against this explanation?

7 Do you agree that evolution happens? Explain why you think this.

8 Suggest why scientists are sometimes reluctant to give up an accepted explanation, even when new data seems to show it is wrong.

Find out about

- ✓ **how new species are formed**

A mutation in a gene controlling fur colour produced tigers with white fur.

What we need here is a bit of variation!

Questions

1 Explain what a mutation is, and how it can happen.

2 What four processes combine to produce a new species?

Key words

- ✓ **mutation**
- ✓ **reproductive isolation**

Charles Darwin's theory of evolution by natural selection predicts that new species will be formed from existing species and that other species will become extinct. These events usually happen slowly over many generations, which is why Darwin was not able to observe them happening. Since Darwin's time scientists have learnt a lot about DNA, and this has helped them to understand how new species form.

Species show variation

We saw earlier that a species is a group of organisms that can breed together to produce fertile offspring. They cannot reproduce successfully with members of different species. All the members of a species are not identical – there is variation.

Mutations cause variation

Suppose that, when DNA is being copied, a mistake is made. This **mutation** could result in a different coloured flower, or spots on an animal's fur. Mutations happen naturally, and they are also caused by some chemicals or ionising radiation.

Mutations produce differences in a species. They are a cause of variation. This is very important for natural selection. Without variation, natural selection could not take place.

Most mutations have no effect on the plant or animal. Mutations that do have an effect are usually harmful. Very, very rarely a mutation causes a change that makes an organism better at surviving. If the mutation is in the organism's sex cells, it can be passed on to its offspring.

Living in an uncertain world

If the environment changes then only some of the population will survive. By natural selection, only individuals with features that make them adapted to the new environment will survive.

Living in splendid isolation

Populations that are isolated from each other have no contact with their neighbours. Organisms will be able to reproduce with other members of their own population, but will never meet organisms from other populations.

Sometimes variation might arise in one population that will prevent the organisms reproducing successfully with those from neighbouring populations, even if they were able to meet. This is called **reproductive isolation**. The isolated population has become a new species.

Variety of life

It has been estimated that there could be over 30 million species on Earth. This huge variety of different animals, plants, fungi, algae, and microorganism species on Earth and the genetic variation within them is called **biodiversity**.

If new species are being formed, does it matter that some become extinct? Isn't extinction just part of life? Twenty-First Century Science put this question to Georgina Mace of the UK Zoological Society.

"It is true that species have always gone extinct. This is a natural process. But the pattern of extinction today is different from what has been recorded in the past.
- The rate of species extinction today is thousands of times higher than in the past.
- Current extinctions are almost all due to humans."

Georgina Mace

Are humans to blame for some extinctions?

In 1598, Dutch sailors arrived on the island of Mauritius in the Indian Ocean. In the wooded areas along the coast they found fat, flightless birds that they called dodos. By 1700, all the dodos were dead. The species had become extinct. The popular belief is that sailors ate them all. But this explanation appears too simple. Written reports from the time suggest that they were not very nice to eat.

What killed the dodos?

Humans may not have eaten dodos. But did they cause their extinction without meaning to? When the sailors arrived, they brought with them rats, cats, and dogs. These may have attacked the dodos' chicks or eaten their eggs. The sailors also cut down trees to make space for their houses. Maybe this took away the dodos' habitat.

So human beings can cause other species to become extinct:
- directly, for example by hunting
- indirectly, for example by taking away their habitat or bringing other species into the habitat.

Dodos were not able to survive the changes in their environment. This is a disaster for any species.

Foxgloves are very poisonous. But they have given us a powerful medicine to treat heart disease.

Monoculture crop production helps to maximise yields and profits but reduces biodiversity.

The number of supermarket plastic bags used is falling, but the total packaging of food and other goods is rising. Large-scale packaging use is not sustainable.

Does extinction matter?

If many species become extinct, there will be less variety on Earth. This variety is very important. For example:

- People depend on other species for many things. Food, fuel, and natural fibres (such as cotton and wool) all come from other species. Some of today's food crops have been developed from wild plants using selective breeding. The wild relatives can still be found, although they are very different to the domesticated varieties. For example, wild potatoes are poisonous and wild sugar cane produces very little sugar. Wild plants will continue to be used for selective breeding to develop new crops.
- Many medicines have come from wild plants and animals. There are probably many other medicines in plants that haven't been found yet.
- Ecosystems that have high biodiversity tend to cope more easily with natural disasters. In a drought, a species with lots of genetic variation is more likely to survive because some individuals will be better adapted to the drier conditions. Some species will not survive at all, becoming locally extinct. But if there are lots of different species in the ecosystem, then some should survive.

Biodiversity and sustainability

Keeping biodiversity is part of using Earth in a sustainable way. **Sustainability** means meeting the needs of people today without damaging Earth for people of the future. The conservation of different species is important if the Earth is going to be a good home for future generations.

Farmers have to grow the food we need to eat but in a sustainable way that will not damage the environment. Farmers prefer to grow crops in large fields, which can be harvested by machines. A single crop of genetically uniform plants will grow at the same rate and be ready to harvest at the same time. This is called a **monoculture**.

One big disadvantage of monoculture is that the crop can be very easily harmed by pests and diseases. This means the farmer often has to use a lot of chemical pesticides to control them. Large fields and the use of pesticides can reduce the biodiversity of the environment, so is not sustainable in the long-term. Large-scale monoculture crop production could also threaten food supply if a natural disaster wiped out the whole crop.

Packaging – the problem

Packaging is useful: it protects food and other goods on their journey from farm or factory to warehouse or shop and then to our homes. But packaging can be an environmental problem. It is one area that could be made more sustainable. We need to think carefully about how we make things and how we dispose of the things we no longer need.

In 2008 an estimated 10.7 million tonnes of packaging was created in the UK. Plastics are light and strong so are used to package about 50% of all goods. All this packaging will end up as waste. In the past the majority of this waste ended up in landfill, a huge hole in the ground that is then covered over and landscaped, hiding the rubbish underground.

Packaging – improving sustainability

The use of **biodegradable** packaging materials such as paper and plant-based plastics, instead of oil-based plastics, would seem to be more sustainable. They should decompose if put in a landfill site, only releasing carbon dioxide that has recently been fixed in photosynthesis. So that it is not just a case of hiding the waste underground. But the bacteria that decompose waste need to have oxygen, and landfill sites often lack oxygen. This oxygen deficiency prevents the breakdown of the waste. Any decay that does occur in low-oxygen conditions produces methane, a powerful greenhouse gas.

One solution to the problem would be to recycle packaging waste. In 2008 the UK recycled 61% of its packaging waste, a massive increase from the 28% recycled in 1997.

It would be even better to reduce the amount of packaging used. This would reduce the amount of resources and energy used to make the original packaging materials and transport them. It would also avoid the need to collect, transport, and recycle the waste, saving more energy and reducing carbon dioxide emissions.

Biodegradable plastics are made from starch and cellulose from plants.

Newspapers dug up after years in a landfill site can still be read.

Questions

1 Explain what is meant by sustainability.

2 Give two reasons why it is important that we do not lose biodiversity of life on Earth.

3 Explain how the use of packaging can be made more sustainable.

Key words

- ✔ **biodiversity**
- ✔ **sustainability**
- ✔ **monoculture**
- ✔ **biodegradable**

Science Explanations

In this module you will consider different explanations for evolution and learn about natural selection. You will learn how living organisms are dependent on their environment and each other for survival, and you will learn about biodiversity and sustainability.

You should know:

- that a species is a group of breeding organisms, producing fertile offspring
- why all species within a food web are dependent on each other
- why organisms compete for resources with other species in the same habitat
- how organisms become extinct if they cannot adapt to environmental change, or if a competitor, predator, or disease-causing organism enters the environment
- that the Sun is the ultimate source of energy for nearly all organisms
- how energy transfers through the ecosystem when organisms are eaten or decay
- how energy is lost from a food chain as heat, waste products, and uneaten parts, limiting the length of the food chain
- how carbon cycles through the environment, including the processes of combustion, respiration, and photosynthesis
- that the nitrogen cycle involves nitrogen fixation, conversion to proteins, excretion, decay, uptake of nitrates by plants, and denitrification
- that life on Earth began 3500 million years ago and evolved from simple living things
- why all individuals are different and that some variation is genetic
- how mutations increase genetic variation and can be passed on to offspring
- how certain characteristics favour the survival of certain individuals in the process of natural selection
- how humans use selective breeding to choose characteristics in plants and animals
- how new species evolve through a combination of mutations, environmental changes, natural selection, and isolation
- that the analysis of DNA and the fossil record provides evidence for evolution
- how the classification of organisms shows their evolutionary relationship
- that Darwin's theory of evolution resulted from observations and creative thought
- that biodiversity includes the variation within and between different species
- how biodiversity ensures sustainability by increasing the stability of ecosystems and is vital for the development of food crops and medicines
- why large-scale monoculture of a single crop does not maintain biodiversity
- that all packaging materials use raw materials, energy for their production and transport, and create pollution; reducing them improves sustainability.

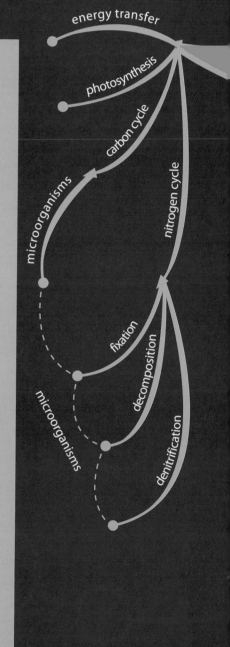

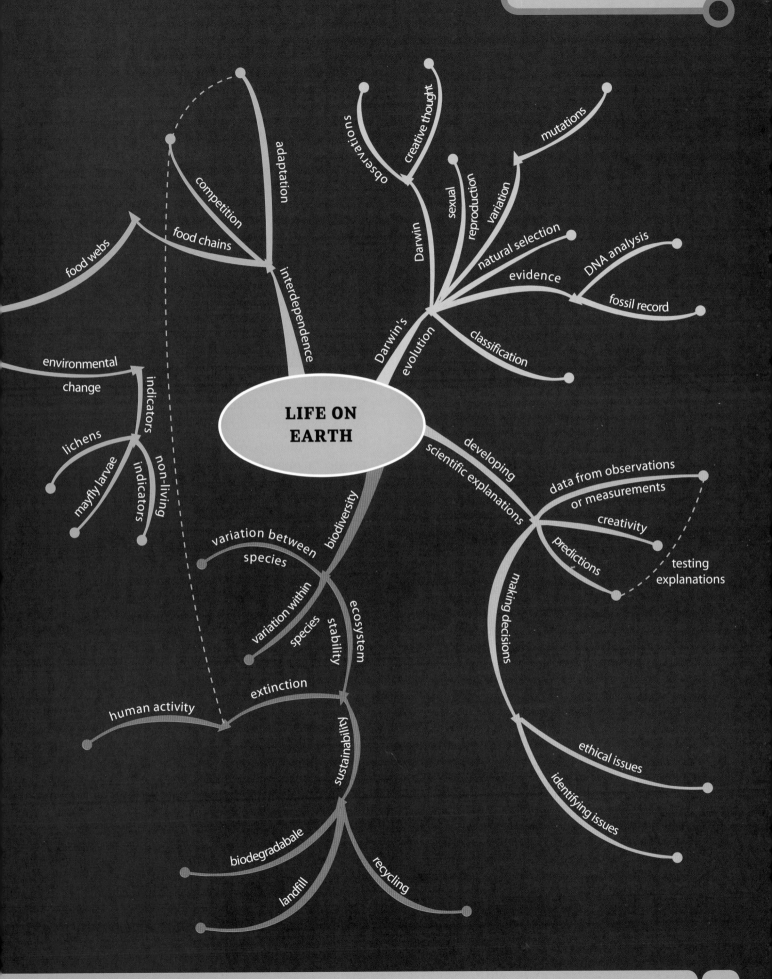

LIFE ON EARTH

adaptation

observations

creative thought

mutations

competition

food chains

Darwin

sexual reproduction

variation

food webs

interdependence

natural selection

evidence

DNA analysis

environmental change

indicators

Darwin's evolution

classification

fossil record

lichens

non-living indicators

mayfly larvae

developing scientific explanations

data from observations or measurements

creativity

biodiversity

predictions

testing explanations

variation between species

variation within species

ecosystem stability

making decisions

human activity

extinction

ethical issues

identifying issues

sustainability

biodegradabale

landfill

recycling

Ideas about Science

Science is about collecting data and using that data creatively to generate explanations. To test whether the explanations are correct, predictions are made and then new experiments or observations are checked against the predictions.

Scientists base their theories and explanations on observations and data. You will need to be able to distinguish between data and explanations and recognise those explanations that involve creative thinking.

Very often, conflicting explanations for the same data are produced. You should be able to suggest why scientists might disagree and to identify the better explanation, giving reasons for your choice. For example, Lamarck thought that characteristics were acquired during life and then passed on to their offspring. By thinking creatively, Darwin produced a new theory; he suggested that organisms evolved due to the process of natural selection. Some scientists disagreed with Darwin because of their personal or religious beliefs.

Scientific explanations can be tested by predictions and comparing the prediction with data obtained from experiments. When the data supports the prediction our confidence in the explanation is increased. For example, we could predict that when a new antibiotic is produced, bacteria will rapidly evolve resistance to the antibiotic by the process of natural selection. If this were to happen it would increase our confidence in the theory of evolution and the process of natural selection.

Science-based technologies improve the quality of our lives. However, science can sometimes have unintended and undesirable consequences. So the benefits of the new technology need to be weighed against the costs. Humans have introduced new animals into ecosystems with the best of intentions but the consequences have sometimes been terrible, such as the introduction of the rabbit and cane toad into Australia.

You will need to be able to suggest examples of unintended impacts of human activity on the environment, and use ideas and data about sustainability to compare the sustainability of different products or processes, for example, using biodegradable products in packaging.

Review Questions

1 The chart shows the evolution of humans (*Homo sapiens*) over the past 7 million years, drawn using evidence from fossils.

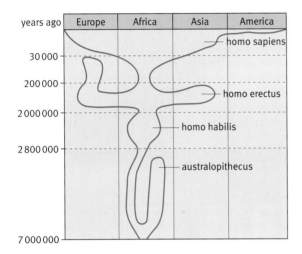

years ago	Europe	Africa	Asia	America
				homo sapiens
30 000				
200 000				homo erectus
2 000 000				
				homo habilis
2 800 000				
				australopithecus
7 000 000				

a Neanderthals are another extinct relative of humans. They did not evolve into *Homo sapiens*. Neanderthals became extinct just over 30 000 years ago. Identify the part of the chart that represents Neanderthals and state which continent they were mainly found in.

b Use the chart to answer these questions.

  i Which of the statements below are true?

    A All the species named on the chart evolved from a common ancestor.

    B *Homo sapiens* appeared before *Homo erectus*.

    C *Australopithicus* evolved from *Homo habilis*.

    D *Homo habilis* spread to more continents than *Homo sapiens*.

    E *Homo erectus* was mainly found in Africa.

  ii Name one species on the chart that is not yet extinct.

2 Look at the food web.

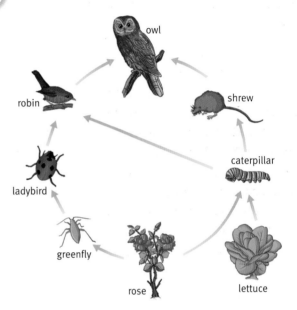

a Explain the effect on the food web of a farmer spraying the caterpillars with an insecticide.

b Write down one food chain found in the food web.

c Explain how energy enters the food chain, how it is transferred along the food chain, and how it is lost from it.

3 Explain the role of microorganisms in the nitrogen cycle.

You may draw a diagram to help your answer.

C3 Chemicals in our lives: Risks and benefits

Why study chemicals in our lives?

You are made up of chemicals, and so is everything around you. Many of the chemicals in the things you buy are natural, others are synthetic. Some chemicals are very good for you, others may be harmful. This module will help you to understand where some of these chemicals come from and why they are so useful.

What you already know

- Clues in the rocks help scientists discover how the Earth has changed.

- The movements of tectonic plates lead to changes in the Earth's surface.

- Chemical reactions rearrange atoms to give products with new properties, which may be either helpful or harmful.

- Alkalis neutralise acids to form salts.

- Crude oil is a valuable source of hydrocarbons.

- Polymerisation produces a wide range of plastics, rubbers and fibres.

- Plasticisers can be used to modify the properties of polymers.

- There are ways to weigh up the risks and benefits of scientific discoveries.

Find out about

- the geological history of Britain, which explains why it is rich in natural resources

- methods chemists use to turn raw materials, such as salt, into many valuable products

- ways to balance the benefits and risks of using chemicals

- the choices people make to ensure they use chemicals safely and sustainably.

The Science

Science can help to explain why Britain has deposits of valuable natural resources including salt and limestone as well as coal, gas and oil. These raw materials have been the basis of a chemical industry that has added to the wealth of the country for over 200 years.

Ideas about Science

Manufactured chemicals bring many benefits – but there are also risks. People are worried that many chemicals have never been thoroughly tested, so the risks are not fully understood. Choices about using chemicals should be based on evidence. The evidence comes from studying all the stages of the life of products.

Find out about

- ✔ **how Britain came into existence as continents moved**
- ✔ **the different climates Britain has experienced**
- ✔ **magnetic clues that geologists use to track continents**

A story of change

The Earth's outer layers (the crust and upper mantle) are divided into a number of **tectonic plates**. Each plate contains dense oceanic crust, often carrying some lighter continental crust on top of it. The plates move because of very slow **convection** currents in the underlying solid mantle.

Movements of the tectonic plates cause oceans to open up slowly between continents in some parts of the world. In other parts of the world, plate movements bring continents together with great force, creating mountain ranges. Most major volcanic eruptions and earthquakes happen at plate boundaries.

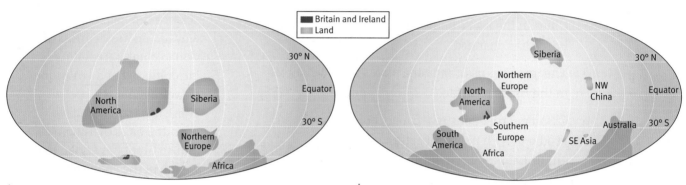

About 450 million years ago, in the Ordovician period, the two parts of the Earth's crust that would one day make up Britain were both south of the Equator. The northern and southern parts were separated by an ocean.

By about 360 million years ago, at the end of the Devonian period, the two parts of Britain collided. The collision between continents created a chain of mountains. The land that would become Britain was at the edge of this chain of mountains in a dry continent.

About 280 million years ago, at the beginning of the Permian period, Britain was just north of the Equator and had desert-like conditions.

About 65 million years ago, dinosaurs (and many other groups of organisms) became extinct. Britain was on the edge of the North Atlantic ocean, just south of where it is today. The Atlantic Ocean was opening up as North America and Europe very slowly moved apart.

Magnetic clues to the past

In the 1950s, the idea that the continents were slowly moving across the Earth's surface was still controversial. Many geologists felt there was not enough evidence to support the idea. At around this time, a group of scientists at Imperial College London, showed that it is possible to track the position north or south of the Equator of a slowly drifting country by studying **magnetic** particles in the rocks.

Many volcanic lavas, and some sediments, contain the mineral magnetite. This mineral gets its name from the magnetic properties of its crystals. Magnetite in lava can be magnetised in a fixed direction once the rock has cooled enough. The magnetisation lines up in the direction of the Earth's magnetic field at that time, rather like iron filings around a bar magnet. The magnetisation of crystals in sediments can line up in a similar way.

Near the Equator, the magnetisation lies horizontally. Nearer to the Poles, the magnetisation is at an angle to the horizontal. So by measuring the angle at which crystals are magnetised in rocks, scientists can work out the **latitude** at which the rock was originally formed.

The measurements were combined with other clues to show that the rocks in Britain had drifted north from a position south of the Equator, over a period of millions of years. The evidence supported the theory of continental drift, and contributed to the development of the theory of plate tectonics. This movement means that Britain has experienced many different climates during its long history. Evidence for these various climates can be found in the different rocks that now make up the country.

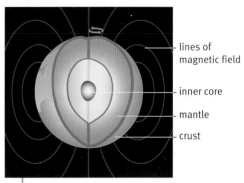

lines of magnetic field

inner core

mantle

crust

A cutaway diagram showing the Earth's magnetic field. The field is parallel to the ground at the Equator, but is more steeply angled nearer the Poles.

Questions

1 What causes continents to move over the surface of the Earth?

2 Do the observations of magnetic minerals made by the scientists at Imperial College support or conflict with the theory of plate tectonics?

3 Suggest evidence that geologists might look for to test the theory that the northern and southern parts of Britain were once on different continents.

4 Movements of the Earth's crust can cause layers of rock to bend and fold. Why might the folding of rocks make it very difficult to interpret the results from measuring the direction of magnetisation in rock samples?

Key words

✓ **tectonic plates**
✓ **magnetism**

Find out about

- ✓ **what geologists can learn by studying rocks**
- ✓ **the origins of some of the rocks in Britain**

Sand dunes in the Namib desert, Namibia. Studying the ripples in today's sand dunes helps to explain the distribution of grain sizes and ripples seen in sandstone rocks.

Fossilised ripples in sandstone on the Maumturk Mountains, County Galway, Ireland.

Key words

- ✓ **sedimentary rocks**
- ✓ **grains**
- ✓ **fossils**
- ✓ **erosion**
- ✓ **evaporation**

Clues in the rocks

Geologists explain the history of the Earth's surface in terms of processes that can be observed today. For example, you can find out about the history of a **sedimentary rock** such as sandstone by looking at the shape and size of the sand **grains** in the rock. The sandstone may have formed from desert sand or river sediments. By comparing the sand grains in the rock with sand grains found in deserts and rivers today, geologists can find out about the conditions when the rock formed. Other clues come from the presence and shape of fossilised ripples in rocks, which may have been produced by the wind or water.

Some sedimentary rocks are rich in **fossils** of plants and animals. Geologists use fossils to put rock layers in order of their ages. This is possible because rocks may contain distinctive fossilised plants and animals from different periods of geological time. Comparing fossils with today's living organisms gives clues about the past environment where the fossilised plants and animals lived.

Different rocks in different climates

There is a rich variety of rocks in Britain, and some are very important economically. A chemical industry based on chlorine grew up by the River Mersey in north-west England because underground salt deposits, coal mines, and limestone quarries were nearby. These provided the raw materials for making chlorine. The salt, coal and limestone formed at different times and in different climates during Britain's long geological history.

Questions

1. Give an example to show how studying a natural process today can tell scientists that processes such as rock formation and mountain building are very slow and take place over millions of years.

2. The chemical industry uses limestone quarried in the Peak District National Park because it is very pure. How do geologists account for the purity of the limestone?

3. Why are fossils mainly found in sedimentary rocks, less commonly in metamorphic rocks, and not at all in igneous rocks?

Limestone from a cavern in the Peak District. It contains fossils of crinoid sea lilies. This limestone formed 350 million years ago. When it formed, the land that would become the Peak District lay below a shallow, warm sea, which was then just south of the Equator. At the time the water was very clear because rivers were bringing in very little sediment. The sea was full of living things. As the plants and animals died they sank to the bottom and formed fossils in the thickening mass of pure limestone.

Coal shale containing a fossil fern. About 280 million years ago, the river deltas in the area that is now Britain got bigger and created swamp land. Tree ferns grew in the swamps. As the plants died, they formed a layer of peat, which was covered by sediment, compressed, heated, and turned to coal. A period of mountain building followed and the rocks in the Peak District were pushed up towards the surface.

geological time period

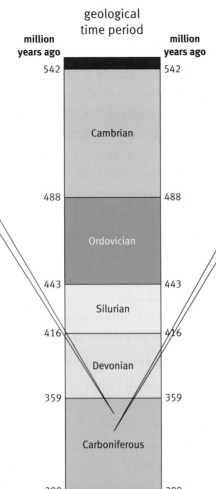

million years ago	geological time period	million years ago
542		542
	Cambrian	
488		488
	Ordovician	
443		443
	Silurian	
416		416
	Devonian	
359		359
	Carboniferous	
299		299
	Permian	
251		251
	Triassic	
199		199
	Jurassic	
145		145
	Cretaceous	
65		65
	Paleogene	
23		23
	Neogene	
2.5		2.5
	Quaternary	

present day

Sandstone in the Peak District. About 310 million years ago, the mountains to the north and east of the Peak District were **eroded** by fast-flowing rivers carrying sediments. Sand and small pebbles were deposited in layers, which were then compacted to form coarse sandstones. In the past, this rock was used to make millstones for grinding corn – it is called millstone grit.

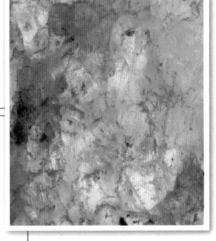

Rock salt mined in Cheshire, consisting mainly of the mineral halite. About 220 million years ago, seawater moved inland and created a chain of shallow salt marshes across land that is now part of Cheshire. Deposits of rock salt formed as the water in the marshes **evaporated**. This rock salt has a red-yellow coloration. The colour is from sand that blew into the salt marshes from surrounding deserts.

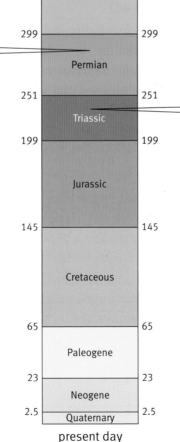

Find out about

- ✓ the uses of salt
- ✓ where salt comes from
- ✓ the methods used to obtain salt

The importance of salt

Salt has played a vital part in human civilisation for thousands of years. Before there were modern ways of keeping food (such as canning, or freezing), salt was the only way to **preserve** meat and fish. After salting, these foods could be kept for a long time, for example, on long sea voyages. Today, salt is still used by the food industry to process and preserve food, and to add flavour. It is also used to treat icy roads in winter and as a source of chemicals such as chlorine.

Sea salt

Salt has been extracted from the sea off the east coast of Essex for over 2000 years. The rainfall in this part of Britain is lower than elsewhere. This means that the concentration of salt in the estuaries and rivers is higher, so less fuel is needed to **evaporate** the water and separate the salt. Small quantities of salt are still obtained in this way for home use.

Large-scale extraction of salt from the sea is only economical on coasts with hot and dry climates. In these places there is no need to burn fuel to separate the salt, because the energy needed to evaporate the water comes from the Sun.

Rock salt

There are two underground salt mines in England – one in North Yorkshire, the other at Winsford in Cheshire. This salt is mainly used to spread on the roads during freezing weather. Adding salt means that ice and snow melt. This is because salty water has a lower freezing point than pure water.

Miners use giant machines to extract rock salt. The rock contains about 90% sodium chloride in the form of the mineral halite. The halite is mixed with insoluble impurities, mainly a reddish clay. The salt used on roads does not need to be pure.

Salt cod in a fish market in Barcelona. In the days before refrigeration, cod caught in the seas off Newfoundland, Iceland, or Norway was salted and dried before being taken by ship to countries such as Portugal and Spain.

Carrying a basket of salt from salt pans near Mahabalipuram in India. Sea water is run into the pans and allowed to evaporate in the hot sunshine. The crystals contain mainly sodium chloride.

Mining rock salt at the Winsford salt mine.

Solution mining

The salt used for the chemical industry in Britain is not mined, it is extracted by pumping water down into the rock. The salt **dissolves** and is carried to the surface in **solution**. The impurities, such as clay, do not dissolve and so stay underground. The solution of salt in water is called **brine**.

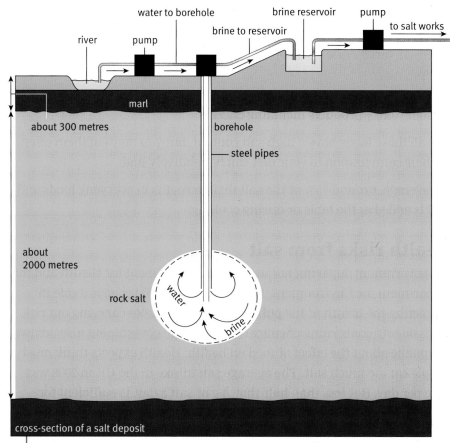

water to borehole
brine reservoir
pump
brine to reservoir
to salt works
river
pump
marl
about 300 metres
borehole
steel pipes
about 2000 metres
rock salt
water
brine
cross-section of a salt deposit

Using water to extract brine from an underground salt deposit.

Salt crystals are recovered from brine by evaporating the water. On such a large scale, the evaporators have to be as efficient as possible to minimise the energy needed to turn water into steam. The salt **crystallises** as the water evaporates. The crystals are separated from the remaining brine by **filtering** or using a **centrifuge**.

Sudden subsidence

Large-scale pumping to extract salt as brine started in Cheshire in about 1870. Uncontrolled pumping for about 60 years created very large underground holes. This led to widespread **subsidence** and flooding of land. From time to time there were disastrous collapses that destroyed buildings. Nowadays, pumping is planned so that the holes in the rock are spaced out and separated by pillars of rock. The rock that is left behind helps to prevent subsidence.

Subsidence caused by salt mining in 1891. The rear of Castle Chambers in Northwich suddenly sank into the ground. The building did not collapse because of its timber frame.

Questions

1 Explain why the east coast of Essex is a good place to extract sea salt.

2 Why is salt for treating roads extracted in a different way from salt used for food and salt used in the chemical industry?

3 Name the solvent and the solute in brine.

4 Outline how you would make some pure salt from rock salt on a small scale in a laboratory.

Salt in food

Salt is sodium chloride. The sodium in salt is an essential part of a healthy diet – but you only need a small amount. Salt is used as a **flavouring** and also enhances other flavours present in food. Because humans need sodium in their diet, they naturally seek out salty-tasting food. The food industry also uses salt to preserve and process food.

The main sources of salt in the diet are:

- cereal products such as bread, chapattis, breakfast cereals, biscuits, and cakes
- processed meat and fish products
- some dairy products including cheese.

Not all these foods have a high salt content, but you may eat them often, so their contribution to your diet can be relatively high.

On average, around 75% of the salt that you eat is in everyday foods and 25% is added at the table or during cooking.

Health risks from salt

UK government departments such as the Department for Health and the Department for Environment, Food, and Rural Affairs have a role in protecting the health of the public. This role includes carrying out risk assessments concerning chemicals in food. The government also advises the public about the effect of food on health. Health experts think most people eat too much salt. The average salt intake in the UK in 2008 was 8.6 g per day. But less than half that, 4 g of salt a day, is sufficient for nearly everyone. Government agencies have been working with health experts, consumers and industry to reduce salt intake to a target of 6 g per day for adults, and less for children.

UK government agencies say that eating too much salt can raise people's blood pressure. This can increase the risk of developing heart disease or having a stroke. People can lower their blood pressure in as little as four weeks by cutting down on salt. Around a third of the population in the UK currently have high blood pressure. When blood pressure goes down, the risk of developing heart disease and stroke goes down too, at any age.

The Scientific Committee on Nutrition is an independent body of experts set up to advise the government. The committee has reviewed over 200 scientific papers from all over the world on the evidence on salt and health. It concluded that 'A reduction in the dietary salt intake of the population would lower the blood pressure risk for the whole population.'

The sodium in salt is an essential part of your diet. It is found in your blood, tears, and nerves. Nerves conduct electrical signals thanks to the presence of salts.

The committee has also concluded that the risk from heart disease associated with high salt consumption is not only confined to those who already have high blood pressure. A large number of people with blood pressure in the normal range are also at risk.

Challenging the salt theory

The European Salt Producers' Association is an industry body representing salt producers across Europe. They have produced a report challenging the **theory** that reducing salt intake brings health benefits for everyone in the population. It suggests that there is no scientific proof for this theory. The report also suggests that a low-sodium diet could be harmful in some cases. To support this point of view, the association refers to two independent reviews of research carried out between 1966 and 2001.

Other scientists also argue that the evidence does not support an approach aimed at reducing salt intake in the whole population, as people with normal to low blood pressure might not benefit.

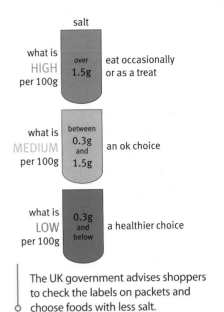

salt

what is **HIGH** per 100g	over 1.5g	eat occasionally or as a treat
what is MEDIUM per 100g	between 0.3g and 1.5g	an ok choice
what is LOW per 100g	0.3g and below	a healthier choice

The UK government advises shoppers to check the labels on packets and choose foods with less salt.

Key words
- ✓ flavouring
- ✓ theory

There is conclusive evidence that moderate sodium reduction lowers blood pressure.

The research does not support a general recommendation to reduce sodium intake.

Cutting salt in the diet may be worthwhile for older people with high blood pressure. For people whose blood pressure is normal, the evidence is not strong enough to justify a general reduction in salt levels.

Questions

1. What are the main sources of salt in the diet?

2. A 25 g packet of crisps contains 0.6 g of salt. Do these crisps have a high, medium, or low salt content?

3. Suggest a reason why it is difficult to investigate scientifically the health risks of different levels of salt in the diet.

4. Suggest reasons why UK government agencies and the European Salt Association might come to different conclusions about the effects of salt levels in the diet on health.

5. Why might a member of the public, who is not an expert, ignore the advice of the government and eat more than the recommended amount of salt?

Find out about

- ✓ uses of alkalis
- ✓ where alkalis used to come from
- ✓ neutralisation of acids with alkalis

Traditional alkalis

Even before large-scale industrialisation, **alkalis** were needed:

- to neutralise acid soils
- to convert fats and oils into soap
- to make glass
- to make chemicals that bind natural dyes to cloth.

Alkalis for making alum

One of the first pure chemicals made in Britain was alum. The biggest use of alum was for dyeing cloth. Alum was needed to help natural dyes cling fast to cloth so that the colours did not fade too quickly during washing.

Alum was made on the north-east coast of Britain, where rock from the cliffs is unusually rich in aluminium compounds. Workers used to roast this rock in open-air fires for many months. Then they tipped the burnt rock into pits of water and stirred the mixture with long wooden poles.

After allowing the waste rock to settle, they ran the solution of soluble chemicals into lead pans. There they boiled the liquid to get rid of much of the water and added an alkali to neutralise acids in the solution. Finally they allowed the solution to cool in wooden casks. When they broke open the casks, found crystals of alum, which they could crush and put into bags for sale.

Some of the alkali used in this process was potash, from the ash of burnt wood. The rest was ammonia, from stale urine. Local people stored urine in wooden pails and this was collected in large barrels on horse-drawn carts. So much urine was needed that it was also brought in by sea from London. On the return journey, the ships delivered the bags of alum to dyers in the south of England.

Alkalis and their reactions

All alkalis are soluble in water – at least to some extent. When they dissolve, they raise the pH of water above 7. Alkalis are important because they **neutralise** acids.

Two very corrosive alkalis are sodium hydroxide and potassium hydroxide. When sodium hydroxide neutralises hydrochloric acid, there is a chemical change that produces sodium chloride. Chemists sometimes call this 'common salt', because it is just one of many

The old way to make soap. Soap was made by boiling up animal fat with the alkali potash. The potash (potassium carbonate) came from the ashes of burnt wood.

Glass is made by melting pure sand (silicon oxide) with lime (calcium oxide) and soda ash (sodium carbonate).

different **salts** produced when acids and alkalis react. This reaction is shown by the **word equation**:

sodium hydroxide + hydrochloric acid $\longrightarrow$ sodium chloride + water

There is a pattern to these reactions:

alkaline hydroxide + acid $\longrightarrow$ salt + water

If the acid used is hydrochloric acid, the salt will be a chloride. If the acid used is sulfuric acid, the salt will be a sulfate. If the acid used is nitric acid, the salt will be a nitrate.

Sodium carbonate and potassium carbonate dissolve in water to form a solution with a pH above 7. They are also alkalis. They fizz when they are mixed with an acid because the reaction produces carbon dioxide as well as a salt and water.

$$\text{potassium carbonate} + \text{sulfuric acid} \longrightarrow \text{potassium sulfate} + \text{carbon dioxide} + \text{water}$$

Again, there is a pattern to the reactions:

alkaline carbonate + acid $\longrightarrow$ salt + carbon dioxide + water

Questions

1 At which stages of the manufacture of alum were the following processes involved? Which of these processes involved chemical reactions to make new chemicals?
  a oxidation
  b dissolving
  c evaporation
  d neutralisation
  e crystallisation.

2 Stale urine contains 2 g of ammonia in 100 cm³ of the liquid. The daily output of a person is about 1500 cm³ of urine.
  a Estimate the mass of ammonia, in tonnes, that could be obtained per person per year (1 tonne = 1000 kg, 1 kg = 1000 g).
  b 3.75 tonnes of ammonia is needed to make 100 tonnes of alum. Estimate the number of people needed to supply the urine for an alum works producing 100 tonnes of alum per year.

3 Name the products of the reactions of:
  a calcium hydroxide with hydrochloric acid
  b potassium hydroxide with sulfuric acid
  c sodium carbonate with nitric acid.

Chalk and limestone contain calcium carbonate. They are insoluble in water. Heating them in a lime kiln, like the traditional one shown here, breaks down the calcium carbonate into calcium oxide. Calcium oxide reacts with water to make calcium hydroxide, which is slightly soluble in water, and can **neutralise** acids, including acids in soils.

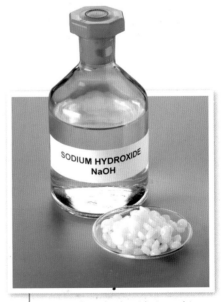

Pure sodium hydroxide is a white solid. It is soluble in water and used in solution as an alkali. Pure sodium hydroxide is very corrosive.

Chemicals from salt – the foul way

Find out about

✓ **how alkalis were first manufactured on a large scale**
✓ **why this was such a polluting process**
✓ **how Parliament began to regulate the chemical industry**

Air pollution from industry in Widnes in 1895.

Making alkali on a large scale

During the industrial revolution in the 1700s, natural sources of alkali became too scarce to meet demand. The shortage of alkali was particularly serious in France, where large amounts of alkali were needed for use in the glass industry.

In 1791, Nicolas Leblanc invented a new process that used chalk or limestone (calcium carbonate), salt (sodium chloride), and coal to make the alkali sodium carbonate. He was awarded a patent by the King of France and given enough money by the Duc d'Orléans to build a plant to operate the process.

This was the time of the French Revolution, and in 1794, the Duc d'Orléans was guillotined. As a result, Leblanc's factory was seized and the patent for his process became public property.

Now others could benefit from Leblanc's invention. A chemical industry began in England based on his process and continued for over 100 years. The industry grew rapidly. The rural areas of Widnes and Runcorn on the banks of the River Mersey were transformed into international centres for new industries based on salt.

The **Leblanc process** was highly polluting. For every tonne of the product sodium carbonate, the process created two tonnes of solid waste. It also released almost a tonne of **hydrogen chloride gas** into the air. This acid gas devastated all the land around. The solid waste was dumped in vast heaps outside the factory, where it slowly gave off a steady stream of toxic **hydrogen sulfide gas**. This gas has a sickening smell of bad eggs. Estimates suggest that by 1891, about 200 hectares of land around Widnes had been buried under an average depth of 3–4 metres of waste. Living and working conditions in the area were appalling.

First steps towards regulating the chemical industry

As industrialisation increased in the 1800s, the British public began to demand action from the government to control pollution. At that time the government was anxious not to restrict the chemical industry because it brought money to the economy and provided jobs. But Parliament did begin to pass laws to regulate working conditions and control pollution from railway engines and factory smoke.

Pollution by the chemical industry became so bad that in 1863, Parliament passed the first of the **Alkali Acts**. This Act set up an Alkali Inspectorate. Inspectors travelled the country to check that at least 95% of acid fumes were removed from the chimneys of chemical factories. The inspectors were scientists. Dressed in Victorian frock coats and top hats, they carried their measuring equipment up long ladders to the top of factory chimneys. There they sampled the smoke and fumes in all weathers.

Tackling the pollution problem

The first response of the Leblanc industry to the Alkali Acts was to dissolve the hydrogen chloride in water. They had no use for the hydrochloric acid that formed, so to begin with they just let it flow through sewers into the local rivers, where it killed all the life in the water.

In 1874, Henry Deacon invented a better way to use the acid gas from the Leblanc process. He found that it was possible to oxidise hydrogen chloride to **chlorine**. Chlorine is one of the elements that make up hydrogen chloride, but it has very different properties to the compound. Hydrogen chloride is corrosive and acidic, while chlorine can be used as a **bleach**.

In Henry Deacon's process, he mixed hydrogen chloride with oxygen and let the two gases flow over a hot catalyst. The products were chlorine and steam. The chlorine could be used to bleach paper and textiles.

The problems of the Leblanc process were eventually solved towards the end of the 1800s, not by government controls, but by developing new methods for manufacturing alkalis. The new processes are still in use today (see H: Chemicals from salt – a better way).

"...they were ruined when they were required to send labouring children to school; they were ruined when inspectors were appointed to look into their works; they were ruined when such inspectors considered it doubtful whether they were quite justified in chopping people up in their machinery; they were utterly undone when it was hinted that perhaps they need not make so much smoke."

HARD TIMES

CHARLES DICKENS

Dickens mocked industrialists' attitudes to new controls when he wrote *Hard Times* in 1854.

Questions

1. Show how hydrogen chloride illustrates the fact that the properties of a chemical compound are very different from the properties of its elements.

2. Draw a diagram to show four molecules of hydrogen chloride, HCl, reacting with a molecule of oxygen to form two molecules of chlorine, Cl_2, and two molecules of steam (water).

3. Why was turning waste hydrogen chloride into chlorine better than dissolving it in water?

4. Suggest reasons why Parliament was slow to bring in laws to control the new chemical industry, despite the serious risks to health and unpleasantness for workers and for people living nearby.

Key words

- ✓ **hydrogen chloride gas**
- ✓ **hydrogen sulfide gas**
- ✓ **chlorine**
- ✓ **bleach**
- ✓ **Leblanc process**
- ✓ **Alkali Acts**

The threat of waterborne disease

Water that is contaminated by sewage can sometimes carry fatal diseases, such as cholera, typhoid, dysentery, and gastroenteritis. The World Health Organisation (WHO) reports that water quality is still a very serious threat to human health. According to the WHO, more than three million people still die each year as a direct result of drinking unsafe water. Waterborne infections cause about 1.7 million of these deaths. Those who die are mainly children in developing countries.

Water treatment with chlorine

Prince Albert, the husband of Queen Victoria, died of typhoid fever in 1861. The poor state of the drains in Windsor Castle may have been to blame. It was only a few years later that it became normal to filter drinking water and then treat it with chlorine to kill **microorganisms**. This process might have saved Prince Albert's life.

A polluted river in Arusha, Tanzania. In many parts of the world, people can't get clean drinking water. Much untreated sewage flows into streams and rivers.

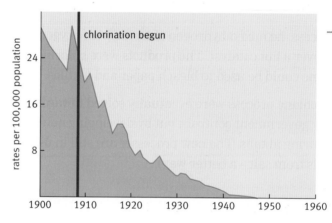

Death rate from typhoid fever in the USA, 1900–1960 (first published in the US Center for Disease Control and Prevention's Summary of Notifiable Diseases 1997).

Chlorination of drinking water in Britain became increasingly common in the early twentieth century. This led to a steep decline in deaths from typhoid. An advantage of using chlorine to treat water is that some of the chemical stays in the water. This means that it can protect against possible contamination in the pipes that carry it from the treatment works to the consumer.

As well as being part of the water purification process, chlorine removes unpleasant tastes and helps to stop microorganisms growing in water storage tanks.

One solution to the problem of water pollution is to use filtration. This device for filtering drinking water is made locally in Cambodia. Where necessary, filtered water can also be treated with chlorine bleach before being used to wash and prepare food.

Risks of water treatment

Chlorination has helped to prevent the diseases associated with drinking water. But some scientists are concerned that there may be side-effects of chlorination. They are worried that chlorination can lead to the formation of a group of chemicals known as trihalomethanes (THMs).

THMs can form when chlorine reacts with naturally found **organic matter** in water, such as fragments of leaves. Some organic matter is found in surface water, such as the lakes and rivers used for drinking water. During the cleaning process, very small amounts of THMs may be formed. When people drink the water, these THMs may be absorbed into their bodies.

There is a suspicion that THMs could lead to some forms of cancer. However, research studies have not found any firm evidence to support this idea. The International Agency for Research on Cancer and the World Health Organisation both say there is not enough evidence to prove any strong link between cancer risks and THMs.

Where organic matter is detected in the raw water to be used for drinking, water companies have devised methods to limit the levels of THMs in the water supply. Ozone gas is used to break down the organic material, then carbon filters can remove it before disinfection by chlorine.

Questions

1 Use the graph to estimate the number of deaths from typhoid in the USA:
   a in 1900, when the population was 76 million
   b in 1940, when the population was 132 million.

2 Water for homes can come from underground aquifers and springs, or it can come from surface sources such as rivers and reservoirs.
   a After chlorination, which type of water is more likely to contain THMs?
   b Suggest reasons why the risk of THMs forming in tap water varies with the time of year.

Dr Harriette Chick (1875–1977) carried out her research at the Lister Institute, a centre for the study of diseases that was then based in London. In 1908, she published the results of her study into the factors affecting how quickly chlorine kills bacteria and viruses in water. Her 'laws of disinfection' helped water companies understand how to use chlorine effectively.

Find out about

- ✓ **the use of electricity to make new chemicals**
- ✓ **the chemicals made by the electrolysis of brine**
- ✓ **the environmental impact of the chemical industry based on salt**

Chemicals from salt

Chlorine is now made on a very large scale. Over 10 million tonnes of chlorine are produced from salt in Europe each year. Today electricity is used to make chlorine from salt. This process is very much cleaner than the old Leblanc process.

The Ineos Chlor chemical plant in Runcorn. Chlorine from the electrolysis of brine is used to make PVC and other chemicals.

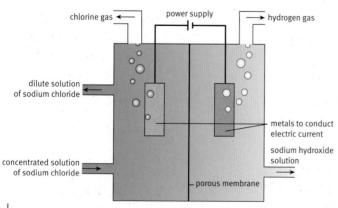

The electrolysis of brine. The porous membrane keeps the chlorine and the sodium hydroxide apart, but doesn't stop the electric current.

Brine is a solution of sodium chloride (NaCl) in water (H_2O). So there are just four elements in the brine, which can be rearranged to make chlorine (Cl_2), sodium hydroxide (NaOH), and hydrogen (H_2).

The chemical changes happen when an electric current flows through the solution. The process is called **electrolysis**.

The chemical changes take place at the surface of the metals that conduct the electric current into, and out of, the solution. The equipment for the electrolysis of brine has to be carefully designed to keep the two main products, chlorine and sodium hydroxide, separate. These two chemicals have to be kept apart because they react with each other when they mix.

Uses of chemicals from salt

The chemicals produced from salt have many uses.

Chlorine	Sodium hydroxide	Hydrogen
• to treat drinking water and waste water • to make bleach. • to make hydrochloric acid • to make plastics including PVC • to make solvents	• to make bleach • to make soap and paper • to process food products • to remove pollutants from water • for chemical processing and products • to make fibres	• to make hydrochloric acid • as a fuel to produce steam

Sodium hydroxide and chlorine react to make bleach.

Chlorine and hydrogen react to make hydrogen chloride gas. The gas dissolves in water to make hydrochloric acid.

Environmental impacts

Manufacturing chemicals from salt by electrolysis needs a lot of energy.

The chemical plant at Runcorn uses as much electricity as a city the size of Liverpool. At the moment, most of the electricity for the electrolysis of brine is generated using fossil fuels. However, the industry is moving towards producing much more of the electricity it needs from renewable sources. It is building a power plant to generate energy by burning household and industrial wastes that can't be recycled.

Until quite recently, the most common system in Europe for the electrolysis of brine used mercury as one of the metals in contact with the solution. Clever design means that chlorine can form in one part of the apparatus, while the flow of mercury circulating in the apparatus allows sodium hydroxide to form in another part. Unfortunately, this method produces products that are contaminated with very tiny amounts of mercury. Also, some of the **toxic** mercury escapes into the environment. As a result, the use of mercury is being phased out.

Increasingly, the industry uses equipment with a sophisticated polymer membrane to keep chlorine and sodium hydroxide apart. This uses less energy for electrolysis, but the sodium hydroxide formed is more dilute so it has to be concentrated by evaporation. Even so, it is the more efficient process.

Changes in the emissions of mercury from European chemical plants for the electrolysis of brine from 1995 to 2008 (published in *Chlorine Industry Review 2008–09*). Some of the mercury ends up in the products, some in waste water, and some in the air.

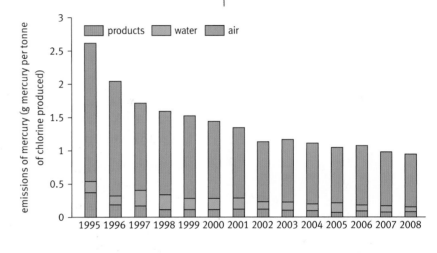

Questions

1 Name the four chemical elements in brine.

2 How does a change in the method used to electrolyse brine account for the fall in the amount of mercury lost to the environment per tonne of chlorine in recent years?

3 What are the reasons for cutting down on the use of fossil fuels to generate electricity for the electrolysis of brine?

Find out about

- why there is a need to check on the safety of a very large number of chemicals
- the European Union's programme for testing
- the problem of persistent and harmful chemicals

Greenpeace activists hold banners reading 'Everyday Chemicals Harm My Sperm!' as they demonstrate in front of the Chancellery in Berlin in 2005.

REACH stands for the Registration, Evaluation and Authorisation of Chemicals.

Untested chemicals

Campaigns by environmental pressure groups, such has the World Wide Fund for Nature (WWF) and Greenpeace, have made many people fearful of **synthetic** (man-made) chemicals. The campaigns have highlighted evidence suggesting that chemicals, such as those found in plastics and pesticides, may cause cancer, or lead to defects in new-born babies.

Most scientists who study toxic chemicals agree that some commonly used synthetic chemicals can be harmful in large doses, but not at the concentrations usually found in people's bodies. Very sensitive chemical tests have typically found only very tiny amounts of these chemicals, less than one part per billion, in human blood. Scientists argue that campaigners are confusing risks with hazards. The chemicals may not be completely safe, but there is no evidence that such tiny traces of them are unsafe.

REACH

Up until 2007, many environmentalists campaigned to try to make the new European Union (EU) laws about chemicals as strict as possible.

European industry produces or uses 30 000 different chemicals a year – a tonne or more of each one. But information about their environmental and health effects is available for only a very small proportion of these compounds. European countries and the USA have been safety-testing all new chemicals since 1981, but this only accounts for about 3% of those in use.

In 2007, the EU introduced the REACH system to collect information about the hazards of chemicals and to assess the risks. REACH switches most of the responsibility for control and safety of chemicals from the authorities to the companies that make them, or use them. Now industry has to manage the risks of chemicals for human health and the environment.

POPs and pollution

There are some synthetic chemicals that everyone agrees are harmful even in very small amounts. These are chemicals that do not break down in the environment for a very long time. This means they can spread widely around the world in air and water. They build up in the fatty tissues of animals, including humans. So they can harm people and wildlife.

This set of chemicals is sometimes called the 'dirty dozen'. They are:

- eight pesticides (two of them are DDT and DDE)
- two types of compounds used in industry (including PCBs)
- two by-products of industrial activity (including dioxins).

All the chemicals in the 'dirty dozen' list are classified as **persistent organic pollutants** (POPs). Many of them are compounds containing chlorine. They are a particular problem for people living in the Arctic, where traditional diets are often high in fat. POPs tend to **accumulate** in fatty tissue of animals, which people then eat.

Experts at a conference in Stockholm in 2001 agreed a Convention to deal with POPs. It became effective in 2004, and about 150 countries have agreed to outlaw the 'dirty dozen' chemicals.

The 10 pesticides and industrial chemicals listed in the Convention have all been banned in Britain for several years. The two other POPs have never been produced intentionally, but may be formed as by-products during combustion of wastes, or in some industrial processes.

Key words

✔ **accumulate**

✔ **synthetic**

✔ **persistent organic pollutants**

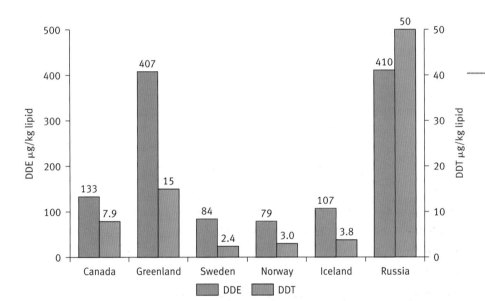

DDT and DDE levels in the blood plasma of mothers in different Arctic regions (published by the Arctic Council Secretariat). The units are micrograms per kilogram of fatty chemicals (lipids). A microgram is a millionth of a gram.

Questions

1 Give a chemical example to explain the difference between a hazard and a risk.

2 Why can't scientists say for sure that small traces of permitted chemicals are completely safe?

3 Manufacturers and importers will pay for most of the cost of REACH. The cost for the whole EU will be about €5 billion over the first 11 years of the testing programme. The EU has a population of about 500 million people.

a Do you think that such a large cost is justified? Explain your answer.

b Is it right that industry should have to organise and pay for the testing? Explain your answer.

4 Suggest reasons why the Stockholm convention allows the use of the insecticide DDT to control mosquitoes in parts of the world where malaria is a serious problem.

PVC is a synthetic polymer. It is strong, easy to mould and quite cheap. It is also hard-wearing, durable, and can be used to make a wide range of products. The stages in the life of PVC products include production, use, and disposal.

Chemicals from raw materials

PVC is made from two chemicals: ethene and chlorine. In the first stage of the process, chlorine and ethene are combined to make a chemical that is commonly called vinyl chloride. This is a hazardous compound because it can cause cancer. PVC manufacturers take great care to make sure that workers are not exposed to this chemical.

Making PVC from chemicals

Vinyl chloride is a liquid made up of small molecules. These small monomer molecules are joined together to make long chains of poly(vinyl chloride) (PVC) by polymerisation. PVC molecules are made up of three different elements: carbon, hydrogen, and chlorine.

The polymer can be supplied as small plastic beads ready for moulding into products.

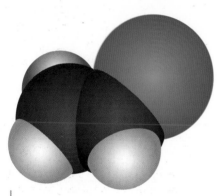

A molecule of vinyl chloride, C_2H_3Cl. Carbon atoms are shown as black, hydrogen atoms white, and the chlorine atom green.

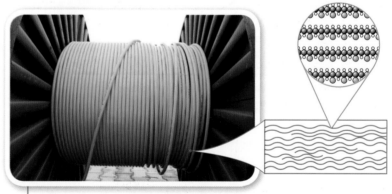

PVC is used to insulate electric cables. PVC is a polymer. The red lines represent the molecules, which are very long chains. These long chains are made up of carbon, hydrogen, and chlorine.

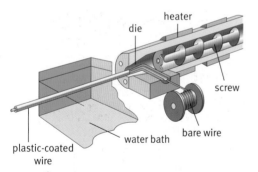

Extrusion is a way of coating copper wire with plastic. The machine heats the polymer to melt it. Then the screw forces the softened plastic though a die so that it coats the wire being pulled through the die at the same time. This process needs energy to heat the plastic and run the machinery. Water is used to cool the mould.

Making products from PVC

PVC granules are sent to factories to be moulded under heat and pressure. For example, the hot plastic can be **extruded** into pipes or blown to make bottles.

Using PVC products

About half of the PVC made each year is used to make pipes. Many of these pipes are underground, carrying drinking water, sewage, and gas. Much of the rest of the PVC produced is used in building, for gutters and window frames. Many commercial signs are made from sheets of PVC.

A softer type of PVC is used to make clothing, garden hoses and insulation for electric wires. PVC film is used for packaging, blood bags, and the bags for intravenous drips.

Disposing of PVC products

Recycling

The best way of getting rid of old PVC products is recycling. It is sometimes possible to grind the waste into pellets. The pellets can be reheated and moulded into new products.

Recycling cuts down the amount of raw materials used to make new PVC. It also reduces the amount of waste from PVC products that have reached the end of their life.

A major problem with recycling PVC is that it is often mixed with other materials. This can make separating, sorting, and recycling difficult and expensive. New methods of recycling are being developed for these mixed sources of PVC.

Energy recovery

Some polymer waste can be burnt. The energy released can be used to generate electricity. This is done in special **incinerators**.

Plastics have to be burnt at a very high temperature to avoid releasing hazardous chemicals. This is a particular problem with PVC, because it produces hydrogen chloride gas when it burns. Acid gases can be removed from the fumes produced by burning before they are released into the air. Burning PVC may also produce toxic dioxins if the conditions in the incinerator are not controlled correctly.

Landfill

Unfortunately, much polymer waste still ends up being tipped into holes in the ground. We call this **landfill**. This really is a waste.

A woman sorting plastic waste in Mumbai, India. She is separating out pieces of PVC and putting them in baskets. She picks the pieces out of a barrel of water. PVC is denser than water so it sinks. Other plastics are less dense than water so they float.

Key words
- extruded
- incinerator
- landfill

Questions

1 What is the raw material needed to make:
   a ethene?
   b chlorine?

2 Which three chemical elements are present in PVC?

3 It might seem better to re-use articles made of PVC rather than recycling them or throwing them away. Why might this be impossible or undesirable?

4 Why is it important that waste incinerators do not release hydrogen chloride into the air?

5 People often campaign when there are plans to build a waste incinerator near to where they live.
   a Suggest risks that the campaigners might be worried about.
   b What is it about the possible risks from burning waste that make people so worried?
   c Suggest arguments that might be used to defend the setting up of a waste incinerator.

Find out about

- ✔ **the chemical used to plasticise PVC**
- ✔ **why plasticisers may be harmful**
- ✔ **what the regulators are doing about the risks**

Worries about plasticisers

Toymakers like to use PVC because it is very versatile:

- it can be either flexible or rigid
- it can be mixed with pigments to give bright colours
- it stands up to rough play
- it is easy to keep clean.

Plasticisers are chemicals that make PVC soft and flexible. The most common plasticisers for PVC are **phthalates**.

Plasticisers are made up of quite small molecules, which can escape from the plastic and dissolve in liquids that are in contact with it. For example, plasticisers can escape from a PVC toy into the saliva of a baby that chews it. They can also **leach** out of the plastic used to make blood bags, or the bags for intravenous drips, and so enter patients' blood.

This child is sliding down a rigid plastic slide into a flexible plastic paddling pool.

Blood from donors is stored in plasticised PVC bags. The tubing carrying the blood to patients is also made of PVC, which is usually plasticised with DEHP.

'Ducki' was displayed for the first time at a New York toy fair in 2010. The toy is advertised as being free of PVC and phthalates. It was designed by a pop-art sculptor who began making toys for his daughter because he was worried about the safety of plastic toys.

Key words

- ✔ **plasticiser**
- ✔ **leach**
- ✔ **phthalates**

Some campaigners argue that phthalates should be banned. They think that a ban is justified because of evidence linking plasticisers with health problems such as cancer, liver problems, and infertility.

Makers of PVC and PVC products point out that phthalates have been in use for over 50 years. They say that, in all that time, there has not been a single known case of anyone ever having been harmed as a result of the use of phthalates.

What the regulators say

Regulators in Europe and the USA are concerned about the possible effects of some plasticisers on young children and new-born babies. Since 2007 the European Union has restricted two common phthalate plasticisers to toys that cannot be placed in the mouth. A third plasticiser (DEHP) has been banned completely from toys.

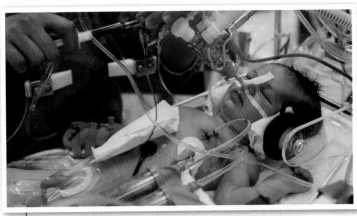

Flexible plastic tubing helps to keep babies alive in intensive care.

Regulators are particularly worried about DEHP because it has been shown to affect the development of the reproductive system and sperm in young male animals. These effects have not been found in human babies but it has not been possible to show that there is no risk. As a precaution, regulators have issued warnings about the use of DEHP in PVC products for medicine.

Plasticisers in medical devices

PVC plasticised with DEHP is used in many medical devices because it has exactly the right combination of properties. It is flexible, strong, and transparent, and keeps its properties at the high temperatures needed for sterilisation (to kill microorganisms) and the low temperatures used for cold storage. It is one of the few materials that has all the right properties and is also affordable.

However, DEHP can leach out of PVC into liquids used to treat patients. Seriously ill people often need treatment for a long time. This can increase their exposure to DEHP. These patients include people who have regular dialysis to treat kidney failure. Others exposed to risk are new-born babies or young children needing repeated blood transfusions.

There are alternatives to DEHP but these are expensive and not always available. This means that if DEHP was banned from medical devices some patients might not receive the treatment they need. Regulators in the USA have told doctors that they should not avoid carrying out medical treatments using plasticised PVC. The regulators say that the risk of not treating a sick patient is far greater than the very small risk from exposure to the plasticiser. However, as a precaution the regulators have recommended that alternatives are used when treating male, new-born babies and women who are known to be pregnant with male fetuses.

Questions

1 Why does it makes sense for regulators to ban the use of the plasticiser DEHP in toys, but only to issue warnings and advice about the use of medical equipment made with PVC softened with the same plasticiser?

2 Why is it so hard to prove that there is no risk when people have fears about possible dangers from a chemical?

3 Some people in medical care are more at risk than others from the possible harm from the plasticiser DEHP.
   a Identify two groups of people who are more at risk and for each group explain why.
   b Give one benefit for patients of using DEHP in medical equipment.
   c If alternatives to DEHP became more affordable and available, how might this affect decisions by hospitals about which material to use?

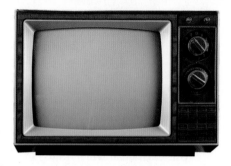

A 1970s TV set. It contains glass, metals, plastics and wood.

The products we buy and use affect the environment. Environmental scientists add up all the effects of a product from cradle to grave. A life cycle assessment can show whether it is better to use a shopping bag made of a natural fibre or a bag made of plastic.

Lives or life cycles

At home, you are surrounded by many different manufactured products – furniture, clothes, carpets, china and glass, televisions, and mobile phones.

The life of each of these products has four distinct phases:

1 The materials are made from natural raw materials.

2 Manufacturers make products from the materials.

3 People use them.

4 People throw them away.

Imagine an old television that was bought in 1970 and thrown away some years ago. It contains glass, metals, plastics, and wood. It is now buried under many tonnes of rubble in a landfill.

The wood will eventually rot because it is **biodegradable**. But the rest of the materials are there forever. This is not sustainable because the materials cannot be used again. The materials had a life, but not a life cycle.

Once the life of a product is over, its materials should go back into another product. This is recycling.

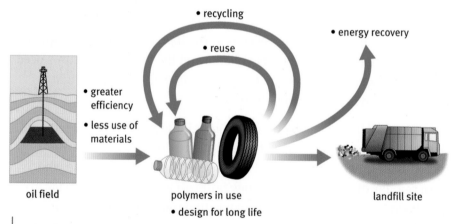

Oil and products from oil, such as polymers, are very valuable. They lose value as they are used and end up as waste. The aim is to slow down the journey of materials from natural resources (cradle) to landfill sites or incinerators (grave).

Life cycle assessment

Manufacturers can assess what happens to the materials in their products. This **life cycle assessment** (LCA) is part of legislation to protect the environment. The aim is to slow the rate at which we use up natural resources that are not renewable. At each stage in the life of a product, raw materials, water, and energy may be used:

- **raw materials** obtained and processed to make useful materials
- **materials** used to make the product
- **energy and water** used in processing and manufacturing

CRADLE

- **energy** needed to **use** the product (eg electricity for a computer)
- **energy** needed to **maintain** the product (eg cleaning, mending)
- **water** and **chemicals** needed to maintain it

USE

- **energy** needed to **dispose** of the product
- **space** needed to dispose of it.

GRAVE

Key words
- ✓ **life cycle assessment**
- ✓ **biodegradable**

An LCA involves collecting data about each stage in the life of a product. The assessment includes the use of materials and water, energy inputs and outputs, and environmental impact. An assessment of this kind can show, for example, whether it is better to make window frames from the traditional material (wood) or the modern material (PVC).

PVC windows need little maintenance. Unlike wooden windows, they are not affected by moisture. Combined with double-glazing, they provide good thermal insulation.

Restoring the paintwork on a wooden window. Wood can rot if it is not properly looked after. It needs regular painting. The advantage of painting is that you can change the colour easily. With double glazing, wood gives good thermal insulation.

Questions

1 Give two reasons why it is not a good idea to put products in landfill once we have used them.

2 Suggest examples of how to slow down the flow of materials from raw materials to waste. Include examples of:
   a re-use
   b recycling
   c recovering energy.

3 Choose a product that has been designed to reduce its impact on the environment.
   a Describe the product.
   b Explain how its environmental impact has been reduced.

Science Explanations

Salt, limestone, coal, gas, and oil have been the basis of the chemical industry for many years. The use of manufactured chemicals has brought many benefits but they are not without risk.

You should know:

- how geologists explain the past history of the surface of the Earth from what they know about processes happening now
- that the parts of ancient continents that now make up Britain have moved over the surface of the Earth as a result of plate tectonics
- how magnetic clues in rocks help geologists to track the very slow movement of the continents
- how processes such as mountain building, erosion, sedimentation, dissolving, and evaporation have led to the formation of valuable minerals
- that clues to the conditions under which rocks were formed come from fossils, shapes of sand grains, and ripples made by flowing water
- why a chemical industry grew up in northwest England
- that salt (sodium chloride) is important for preserving and processing food, as a source of chemicals, and to treat roads in winter
- that salt comes from the sea or from underground salt deposits
- why the methods used to obtain salt may depend on how it is to be used
- why extracting salt may have an impact on the environment
- why alkalis were needed in pre-industrial times
- that the traditional sources of alkali included burnt wood or stale urine
- that examples of alkalis include soluble metal hydroxides and metal carbonates
- that alkalis neutralise acids to make salts
- how industrialisation led to a shortage of alkali in the 19th century
- why the first process to meet a growing demand for alkali was highly polluting
- how the pollution problems of the old process were reduced by producing useful chemicals such as chlorine
- why using chlorine to kill microorganisms in domestic water supplies has made a major contribution to public health
- that electrolysis is the process now used to make new chemicals, such as sodium hydroxide and chlorine and hydrogen, from salt solution
- that PVC is a polymer containing chlorine
- how the properties of PVC can be altered by plasticisers.

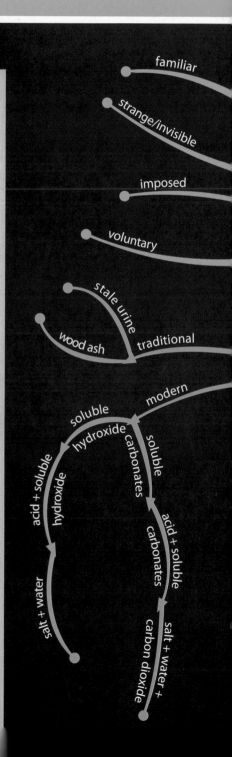

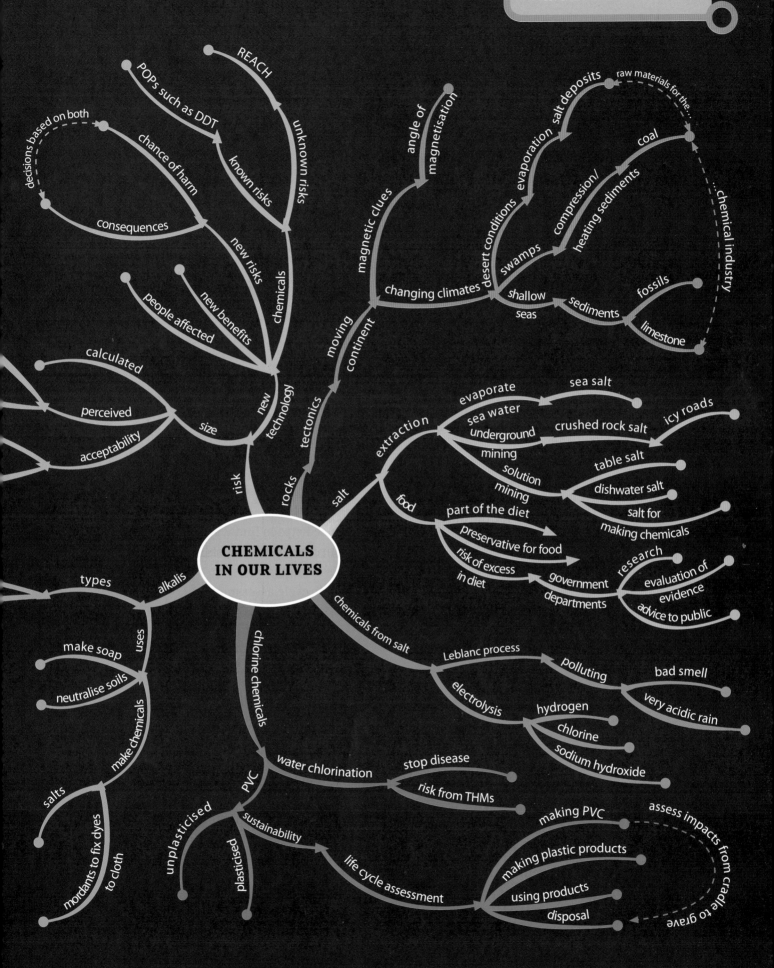

CHEMICALS IN OUR LIVES

REACH

POPs such as DDT

decisions based on both

chance of harm

unknown risks

known risks

consequences

new risks

people affected

new benefits

new chemicals

calculated

perceived

acceptability

size

new technology

risk

rocks

tectonics

moving continent

magnetic clues

angle of magnetisation

changing climates

desert conditions

swamps

shallow seas

evaporation

salt deposits

raw materials for the…

coal

compression/ heating sediments

sediments

limestone

fossils

…chemical industry

salt

extraction

evaporate

sea water

underground mining

solution mining

sea salt

crushed rock salt

icy roads

table salt

dishwater salt

salt for making chemicals

food

part of the diet

preservative for food

risk of excess in diet

government departments

research

evaluation of evidence

advice to public

chemicals from salt

Leblanc process

electrolysis

polluting

bad smell

very acidic rain

hydrogen

chlorine

sodium hydroxide

types

uses

alkalis

make soap

neutralise soils

make chemicals

salts

mordants to fix dyes to cloth

chlorine chemicals

PVC

water chlorination

stop disease

risk from THMs

unplasticised

sustainability

plasticised

life cycle assessment

making PVC

making plastic products

using products

disposal

assess impacts from cradle to grave

Ideas about Science

Scientists seek explanations to account for their findings, such as the data collected by studying rocks. You should be able to:

- explain how magnetic data and other clues in rocks support the theory that the continents have moved.

New technologies and processes based on scientific advances sometimes introduce new risks. Some people are worried about the effects arising from the use of chemicals. You should be able to:

- explain why nothing is completely safe
- identify examples of risks that arise from the use of chemicals
- interpret information on the size of risks, presented in different ways
- describe ways of reducing risks from hazardous chemicals
- discuss a given risk, taking into account both the chances of it happening and the consequences if it did
- identify risks and benefits, for the different individuals and groups involved, arising from uses of chemicals
- suggest why people accept (or reject) the risk of a certain activity, for example, eating a diet with more salt than is recommended
- recognise that people's perception of the size of a risk is often very different from the scientific assessment of the risk
- illustrate the idea that people tend to overestimate the risk of unfamiliar things and things that have an invisible effect.

Governments and public bodies assess what level of risk is acceptable. Treaties, regulations, and laws control scientific research and the applications of science. The decision to regulate may be controversial, especially if those most at risk are not those who benefit. You should understand that governments and regulators are responding to concerns that:

- many people are putting their health at risk by eating too much salt
- there are possible disadvantages of chlorinating drinking water, including possible health problems
- some toxic chemicals persist in the environment – they can be carried over large distances and may accumulate in food and human tissues
- the plasticisers added to PVC can leach out from the plastic into the surroundings where they may have harmful effects.

Science helps to find ways of using natural resources in a more sustainable way. You should understand that:

- a life cycle assessment (LCA) tests:
  - a material's fitness for purpose
  - the effects of using material products from production to final disposal.

- an LCA involves consideration of the use of resources including water, the energy input or output, and the environmental impact, of each of these stages:
  - making materials from natural raw materials
  - making useful products from materials
  - using the products
  - disposing of the products.

When given appropriate information from an LCA, you should be able to compare and evaluate the use of different materials for the same job.

Review Questions

1 a The table shows the mass of salt in different foods.

Food	Mass of salt in 100 g of the food (g)
White bread	1.2
Cornflakes	1.8
Ham	3.1
Crisps	2.5
Chocolate muffin	1.7

  i Which food in the table contains the most salt in 100 g?

  ii Emma eats a 50 g packet of crisps. Ben eats a 100 g chocolate muffin. Who has eaten more salt? Show how you work out your answer.

 b Health experts recommend that adults should eat no more than 6 g of salt each day.

  i Identify two risks linked to eating too much salt.

  ii Suggest why some people eat more than 6 g of salt each day, even though they know about the risks of eating too much salt.

2 a In 1991, there was an outbreak of cholera in South America. The table shows data from the time of the outbreak.

Village	Number of people who caught cholera	Did the village add chlorine to its drinking water?
A	0	no
B	27	no
C	1	yes
D	31	no
E	42	no

Cholera is a disease that is caused by a microorganism. It spreads through contaminated water. Vincent suggests an explanation for the data. He says that adding chlorine to water kills the microorganism that causes cholera.

  i Identify the data that is accounted for by this explanation.

  ii Identify the data that conflicts with the explanation.

 b Describe one possible disadvantage of adding chlorine to water.

 c Give the name of the process by which chlorine is produced from salt (sodium chloride).

3 Sulfuric acid is an acid in acid rain. Sulfuric acid damages limestone buildings. Limestone is mainly calcium carbonate.

Write a word equation for the reaction of sulfuric acid with calcium carbonate.

4 Match each piece of evidence to one or more pieces of information that the evidence can provide.

Evidence	Information
shape of the grains in sandstone	latitude at which the rocks were originally formed
angle at which crystals are magnetised in the rock	whether the rock was formed under the sea
presence of shell fragments	age of the rock
types of fossilised living organisms in the rock	whether the rock was formed from materials from deserts or riverbeds.

P3 Sustainable energy

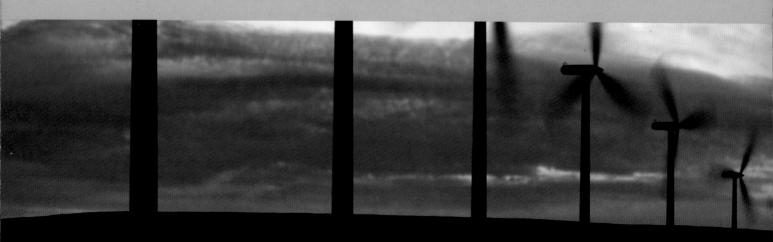

Why study sustainable energy?

Energy supply is one of the major issues that society must address in the immediate future. To make decisions about energy use, you need to understand the figures and calculations behind the headlines. Electricity supplies many of our energy needs. Most of us take electricity for granted. But today's power stations are becoming old and will soon need replacing. How should we generate electricity in the future? Can we reduce our impact on the environment without reducing our quality of life?

What you already know

- When energy is transferred the total amount of energy remains constant.

- Whenever energy is transferred some of it spreads out into the surroundings.

- Electricity is a useful way of transferring energy long distances.

- Electricity is generated in power stations using a variety of energy sources.

- Electric current transfers energy from the power supply to devices in the circuit.

- The higher the voltage of the power supply the more energy the current transfers.

Find out about

- how much energy we use, as individuals, as a country, and across the world

- how we could use energy more efficiently

- how electricity is generated in a power station

- the choices for generating electricity in the future.

The Science

Energy cannot be created or destroyed, but whenever we use it some is lost to the surroundings. It is difficult to recover to use again. We need to find ways of using energy more efficiently. Most UK electricity is generated by burning gas and coal to drive generators. To choose the right generation methods for the future you need to know about the advantages and drawbacks of each.

Ideas about Science

Nothing can be completely safe; there are different risks associated with using each energy source. But who should be making the decisions? How can you have your say?

Find out about

- ✓ energy sources
- ✓ why we need to be concerned about energy supplies

Key words

- ✓ conserved
- ✓ energy source
- ✓ primary source
- ✓ secondary source
- ✓ fossil fuel
- ✓ biofuel
- ✓ pollutant
- ✓ sustainable

People need energy to keep them alive, warm, and able to move. Over time, demand for energy has grown. There are more people now than at any time in the past, and the population continues to expand. Modern transport, buildings, possessions, and communications need more energy than ever before. People travel further and faster and have different expectations about their lifestyle. Understanding about the energy sources available is important when making choices.

Modern living can demand large amounts of energy.

Energy sources

Energy is **conserved**, meaning it can neither be created nor destroyed. The energy we use for heat, movement, and light must all come from an **energy source**. For example, we can release energy by burning fuels, or we can use energy carried by radiation from the Sun.

A **primary energy source** is one that is found or occurs naturally. Examples include fuels such as coal, oil, natural gas, and wood. Wind, waves, and sunlight are also primary energy sources.

Nowadays, much of our energy use involves electricity. Electricity is a **secondary energy source**. It must be generated using a primary source.

Nations and individual people pay for the energy they use. The price is related to the amount of fuel burned, and to the cost of distributing energy to the users. Energy bills are a major expense for most households and for the country as a whole.

Most UK electricity is generated by burning fossil fuels.

Fuels

A **fossil fuel** is one that has built up over millions of years by the decay of plant and animal remains. Coal, petroleum, and natural gas are all fossil fuels. We are using them up far more rapidly than they can form. The table shows how long the world's petroleum supplies will last.

Country	Number of years' supply left (from 2010)
Saudi Arabia	70
Canada	147
Iran	93
Iraq	148
Kuwait	108
United Arab Emirates	91
Venezuela	86
Russia	15
Libya	64
Nigeria	39

Most of the world's petroleum is found in these ten countries.

Petroleum is an oily mixture of solid, liquid, and gas. Petrol, oil, and diesel fuel are all made from petroleum.

When they burn, fossil fuels produce carbon dioxide (CO_2) and other **pollutants**, such as carbon particles. The amount of CO_2 in the atmosphere has risen over the past two hundred years and is affecting the Earth's climate.

A **biofuel** is one that has recently come from living material (biomass). Wood, straw, sewage, and sugar are all used as biofuels. Like fossil fuels, biofuels produce CO_2 when they burn. Unlike fossil fuels, biofuels are produced quickly.

Nuclear fuel releases energy without burning so it does not make CO_2. Nuclear fuels are not found in the UK so any that we use must be imported.

Can wind power supply enough energy?

Sustainable energy

Our current use of energy sources is not **sustainable** and cannot continue indefinitely. Some sources are running out, and some of our energy use is damaging the environment. Governments and individuals must decide how to reduce demand for energy and plan which energy sources to use in the future. To decide what to do, we need to know some facts about the amount of energy we use.

Questions

1 Write down three things that you do during a day that use:
   a a primary energy source
   b a secondary energy source.

2 Suggest at least two reasons for reducing our use of fossil fuels.

We can calculate how much energy appliances use and how much they cost to run. This information can help us make sense of our electricity bills.

Measuring energy

The energy used by an electrical appliance can be measured with a meter. This energy depends on:

- the power rating of the appliance
- the time it is on for.

Power is the rate at which energy is used. An appliance with a rating of 1 **watt** (1 W) uses 1 **joule** of energy (1 J) every second. But 1 watt is a very small power, 1 second is a very short time, and 1 joule is a tiny amount of energy.

Many domestic appliances have powers of a few kilowatts. One kilowatt is one thousand watts. For domestic appliances it is more convenient to use the **kilowatt-hour** as the unit of energy. It is the energy used by a 1 kW appliance switched on for one hour.

1 kWh = 3 600 000 J

energy used	=	power rating	×	time
(joule, J)(watt, W)		(second, s)		
(kilowatt-hour, kWh)		(kilowatt, kW)		(hour, h)

Paying for energy

On an electricity bill, '1 unit' means 1 kilowatt-hour. To find the cost of the energy, multiply the number of units by the cost of one unit.

cost of energy = number of units used × price of one unit

(pence, p)   (kilowatt-hours, kWh)   (pence per kilowatt-hour, p per kWh)

The electricity meter in your home measures the number of kilowatt-hours of electrical energy that you buy.

Meter: 326565		Tariff: **Domestic**	
cost of energy	number of units used	unit charges	total
		first 227 at 13.25p	£30.08
13.25	2213	next 661 at 7.88p	£52.08
			£82.16

Worked example

Working out the cost

A 3 kW immersion heater heats water for a bath in 2 hours. Electricity costs 11p per unit.

Energy used = 3 kW × 2 h = 6 kWh

Cost = 6 kWh × 11 p/kWh

= 66 p

Power, current, and voltage

When an electrical appliance is switched on, electric **current** passes through it and energy is transferred to the appliance and its surroundings. The power is related to the current and to the **voltage** of the electricity supply.

$$\begin{array}{ccccc}
\text{power} & = & \text{voltage} & \times & \text{current} \\
\text{(watt, W)} & & \text{(volt, V)} & & \text{(amp, A)}
\end{array}$$

In the UK, the mains electricity voltage is 230V. For a battery-operated device, the voltage is usually just a few volts.

Appliances that include motors, or are used for heating, usually need quite large currents. They have high power ratings and are expensive to run.

Every electrical appliance has a power rating in watts or kilowatts. This tells you how much energy it uses each second when it is switched on.

Worked example

Working out the current

A hair dryer has a power rating of 700 W.

$$\text{power} = \text{voltage} \times \text{current}$$

For all UK mains appliances the operating voltage is 230 V.

$$700\,\text{W} = 230\,\text{V} \times \text{current}$$

Divide both sides by 230 V to find the current.

$$\text{current} = \frac{700\,\text{W}}{230\,\text{V}}$$

$$\textbf{current} = \textbf{3.0 A}$$

Key words

- ✔ **power**
- ✔ **watt**
- ✔ **joule**
- ✔ **kilowatt-hour**
- ✔ **current**
- ✔ **voltage**

Questions

1   Look carefully at each of the tasks in the table on the right that use electricity, and put them in the order you think they would come – from the cheapest to the most expensive. Then calculate the number of kilowatt-hours that each involves and see if your estimate of the cost was correct.

2   Electricity costs 10p per unit. Calculate how much it would cost to use a fan heater for 2 hours.

3   What is the power of a mains appliance that needs a current of 3 A to make it run?

Task	Appliance used	Power rating (W)	Time for which it is on
watch television for the evening	television	300	5 hours
dry your hair	hairdryer	700	5 minutes
make a pot of tea	electric kettle	2000	4 minutes
write a homework assignment	computer	250	2 hours
listen to music	mp3 player	0.2	2 hours
heat your bedroom while you do your homework	electric fan heater	1500	2 hours
wash a load of dirty clothes	washing machine	1850	$1\frac{1}{2}$ hours
play a game	games console	190	1 hour

Find out about

- ✔ the energy needed for our daily lives

We use far more energy in a day than is accounted for by the electricity bill at the end of the month. How much energy do you use in a day?

Heating and cooking

Most of the energy you use at home is probably supplied by electricity. For some tasks you might use another source such as gas or oil, but the energy needed will be the same.

Key words

- ✔ passenger-kilometre
- ✔ energy cost

Task	Energy (kWh)
Bath (about 100 litres of hot water)	5
Shower (about 30 litres of hot water)	1.4
Gas cooker (for 1 hour)	1.5
Room heater (eg radiator) (for 1 hour)	1
Patio heater (for 1 hour)	15

Transport

Different means of transport use different amounts of energy.

Burning 1 litre of petrol in a car releases about 10 kWh of energy. An economical car can travel about 10 miles (16 km) on 1 litre of fuel, so each mile needs about 1 kWh, which is about 0.6 kWh per km.

Travelling alone in a car uses much more energy than sharing public transport. So figures for public transport are based on having a full vehicle. The table lists energy per **passenger-kilometre**, which is each passenger's share of the energy used to travel 1 km.

A large, less economical car needs about 1.3 kWh per mile (about 0.8 kWh per km).

Food and drink

Growing and producing food uses energy. Some of that energy is stored in the food and passes on to you when you eat it. The table lists the energy needed to produce some fresh foods. Processed foods use more energy.

Transport	Energy per passenger-km (kWh / passenger-km)
Bus	0.19
Train	0.06
Aircraft	0.51
Boat	0.57

Food	Energy (kWh) for production
1 egg	0.5
1 pint of milk	0.8
50 g cheese	0.8
100 g meat (eg beef, chicken, pork)	4
100 g fruit or vegetables	0.5

Other stuff

Everything you use has an energy cost. The table lists the energy needed to make and transport some items that you might buy or use.

Item	Energy (kWh)
Drinks can	0.6
Plastic bottle	0.7
AA battery	1.4
Magazine	1.0
Computer	1800

To find the daily energy cost of an item that you keep for more than a day:

$$\text{Daily energy use (kWh per day)} = \frac{\text{energy needed to produce the item (kWh)}}{\text{number of days that the item lasts}}$$

An average person in the UK uses about 160 litres of clean water per day. The energy needed to treat and distribute this amount of water is the **energy cost** of the water. In the UK the energy cost of our daily water use is about 0.4 kWh.

Worked example

Calculating the energy used

Two people travel 5 km to school in the car.

energy used (kWh) = distance travelled (miles) × energy per km (kWh per km)

energy used = 5 miles × 0.6 kWh per km = **3 kWh**

If they had travelled by bus then

energy used (kWh) = distance travelled (km) × number of passengers
× energy per passenger-km (kWh per passenger-km)

energy used = 5 km × 2 passengers × 0.19 kWh per passenger-km

energy used = **1.9 kWh**

On average each person in the UK throws away 400 g of packaging per day with an energy cost of about 4 kWh.

Questions

1 An aircraft flies from London to New York and back, a total of 5586 km. Look at the information about energy used for transport.
   a If the aircraft carries 500 people what is the total energy used?
   b Explain why the energy cost would be more than half your answer to part a if there were only 250 passengers.

2 Suggest reasons why producing processed food needs more energy than fresh food.

3 Someone wants to change their lifestyle and use less energy. Use information from these pages to suggest what they might do.

Find out about

- ✓ **how public services and activities contribute to our energy use**
- ✓ **how people's daily energy use varies between countries**

In the UK, the average daily energy use per person is about 110 kWh per person per day. A diary of your own energy use will probably give a much lower figure. Where is the rest of the energy used? Are we using more than our 'fair share'?

The bigger picture

We can calculate the average energy used per person in the UK by sharing the total between everybody in the country.

Building homes uses about 1 kWh per person per day.

Building and maintaining roads uses about 2 kWh per person per day.

Supermarkets use about 0.5 kWh per person per day.

The armed forces work on behalf of everyone in the country. Our share of their energy use is about 4 kWh per person per day.

Computer servers are at the core of many businesses and at the heart of the internet. They need energy to drive the computers and even more energy to cool them. Servers across the UK use about 0.5 kWh per person per day.

Global issues

The map shows the average energy use per person in different parts of the world.

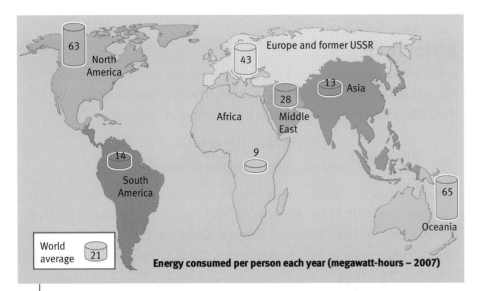

As countries become industrialised, living standards rise and energy use increases. Energy use in India and China is now growing especially fast.

Country	Average daily energy use (kWh per person per day)	GDP per person ($)
Australia	190	47 000
China	50	3 000
Denmark	120	62 000
France	140	45 000
India	20	1 000
Japan	130	38 000
Kuwait	300	54 000
Mexico	60	10 000
Poland	80	14 000
Turkey	40	10 000
UK	110	44 000
USA	250	48 000

Energy use per person and GDP per person 2007 source: World Bank.

The table shows the average daily energy use for people in various countries. The figures for gross domestic product (GDP) per head indicate how rich a country is.

In general, richer countries use more energy per person than poorer ones. If people have more money they can buy more goods, live in larger, more comfortable houses, and travel more. All these things use energy.

But the figures for daily energy use still do not tell the whole story, because they only include energy used within each country. They do not include energy used to make imported goods. To take account of all the UK imports, we should add about 40 kWh per person per day to the UK energy figure.

These jeans are made in China. The energy used to grow the cotton, weave the cloth, and make the garment contributes to the average energy use in China, not the UK.

Questions

1 Suggest at least two more energy-using activities that are shared between everyone in the UK.

2 Plot a scatter graph of daily energy use against GDP for the countries listed in the table. Comment on any trend shown by your graph.

3 'Our energy use is part of a global problem. We should be part of the solution.' Do you agree? Write a paragraph giving reasons for your views.

Find out about

- ✔ **what is meant by 'efficiency'**
- ✔ **how to use a Sankey diagram to show energy transfer**

The CFL on the left needs less energy than the filament lamp on the right to produce the same light output per second. It is more efficient.

We can use less energy by switching off appliances. But we should also use appliances that don't waste energy – appliances that are more efficient.

Efficiency

In electrical appliances, only some of the energy supplied ends up where it is wanted and in the form it is wanted. The rest is wasted, usually as heat. The **efficiency** of an appliance is defined as follows:

$$\text{efficiency} = \frac{\text{energy usefully transferred}}{\text{energy supplied to the appliance}} \times 100\%$$

Until 2009 most UK homes were lit by filament lamps. In a filament lamp, less than one tenth of the energy supplied is carried away as light. The rest is wasted as heat. Filament lamps are now banned in the European Union. They are being replaced by more efficient designs. These include CFLs (compact fluorescent lamps), halogen lamps, and LEDs (light-emitting diodes).

Worked example

Calculating efficiency

An 600 W electric motor is used to lift a load. In one minute the load gains 18 000 J of gravitational potential energy. How efficient is the motor?

Calculate the energy supplied: energy = power × time

$$\text{energy} = 600 \, \text{W} \times 60 \, \text{s} = 36\,000 \, \text{J}$$

Calculate the efficiency:

$$\text{efficiency} = \frac{\text{energy usefully transferred}}{\text{energy supplied to the motor}} \times 100\%$$

$$\text{efficiency} = \frac{18\,000 \, \text{J}}{36\,000 \, \text{J}} \times 100\%$$

$$\text{efficiency} = 50\,\%$$

Question

1 A CFL rated 20 W gives a light output of 11 W.
   a How much energy is used each second by the lamp?
   b How much of this energy is useful?
   c Calculate the efficiency of the lamp.

Sankey diagrams

In a **Sankey diagram**, branching arrows show how energy is transferred. Their width indicates the amount of energy. The total width stays the same because energy cannot be lost or gained overall.

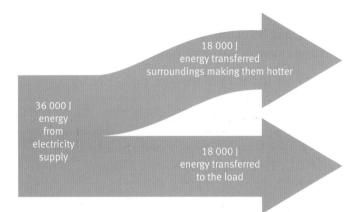

36 000 J energy from electricity supply

18 000 J energy transferred surroundings making them hotter

18 000 J energy transferred to the load

The Sankey diagram for the motor lifting a load with efficiency of 50%.

Electrical appliances are labelled with an efficiency rating to help people choose which to buy.

Questions

2 A filament light bulb rated at 100 W gives a similar output to the 20 W CFL lamp in question 1.
   a Suggest why filament lamps should no longer be sold.
   b Draw Sankey diagrams for both lamps to illustrate your answer.

3 An electric kettle rated 3 kW takes 3 minutes to heat some water. 50 000 J of energy is transferred to the surroundings.
   a How much energy does the electricity supply deliver to the kettle?
   b How much useful energy is transferred to the water?
   c What is the efficiency of the kettle?
   d Draw a labelled Sankey diagram for the kettle.

4 A fundamental law of physics is that energy is always conserved. Energy cannot be created or destroyed. Explain how a Sankey diagram shows this.

5 Explain how using energy-efficient appliances is an advantage to the person and also to the country.

Key words

✓ **efficiency**
✓ **Sankey diagram**

Find out about

- ✔ **the main types of fuel used in the UK**
- ✔ **the advantages of using electricity**
- ✔ **how a magnet moving near a coil can generate an electric current**
- ✔ **how this is used to generate electricity on a large scale**

What fuel do we use and where?

Each year the UK government publishes information about the country's energy use. They refer to three main energy sources: electricity, natural gas, and petroleum (petroleum includes petrol, diesel, and all other fuels made from oil). Other fuels, including solid fuels such as coal and wood, are all grouped together as 'other'.

The main UK energy users are industry, transport, and domestic. Each has a different pattern of fuel use.

Energy use by UK industry 2008.			
Electricity	Gas	Petroleum	Other
33%	38%	21%	8%

Energy use by UK transport 2008.			
Electricity	Gas	Petroleum	Other
1%	0%	97%	2%

UK domestic energy use 2008.			
Electricity	Gas	Petroleum	Other
22%	69%	7%	2%

Questions

1 Write down four tasks that you normally do, using electricity. For each one say whether you could easily use a primary fuel instead.

2 Draw pie charts to show energy use by UK industry in 2008. Draw similar charts for transport and for domestic use.

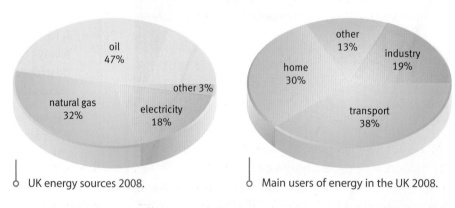

UK energy sources 2008.

Main users of energy in the UK 2008.

In our homes, we might use gas, oil, or other fuels for heating and cooking. For almost everything else we use a secondary source – electricity. Electricity can be used for many different tasks and it is easy to distribute using cables and wires.

Generating electricity

Nowadays most of us in Britain take a mains electricity supply for granted. But it was only in May 2003 that Cym Brefi in mid-Wales became the last village in Britain to get a mains electricity supply.

Electricity reaches last village in Britain

Not having a mains electricity supply does not mean you cannot use electrical appliances. Many can be run from batteries. But this works only for relatively low-power devices. For others, you might use a diesel-powered **generator**.

This is how the inhabitants of Cym Brefi ran their washing machines and vacuum cleaners before they got mains electricity. But generators are noisy, and each 'unit of electricity' is much more expensive than from the mains. So they can only be run for a short time.

A simple generator

Generators work on the principle of **electromagnetic induction**. This phenomenon, which does so much to make our lives comfortable and convenient, was discovered in the 1830s by Michael Faraday.

One way to generate a current is to move a magnet into, or out of, a coil. The movement of the magnet causes an induced voltage across the ends of the coil. 'Induced' means that it is caused by something else – in this case, the movement of the magnet. The coil, for a brief time, is like a small battery. If the coil is part of a complete circuit, this induced voltage makes a current flow.

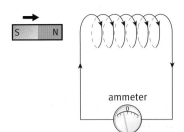

1. While the bar magnet is moving into the coil, there is a small reading on the sensitive ammeter.

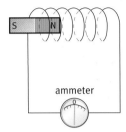

2. There is no current while the magnet is stationary inside the coil.

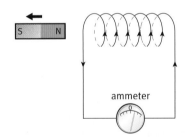

3. While the magnet is being removed from the coil, there is again a small current, but now in the opposite direction.

Moving a magnet into, or out of, a coil generates a current.

Key words
- ✓ **generator**
- ✓ **electromagnetic induction**
- ✓ **alternating current (a.c)**

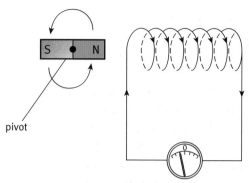

rotate magnet

pivot

needle moves
back and forth

A rotating magnet generates an
alternating current in the coil.

Continuous current

If a magnet is repeatedly moved in and out of a coil, or rotated close to a coil, a continuous to-and-fro current (an **alternating current, a.c.**) can be generated. This is what happens inside a shake torch, a wind-up radio, some bicycle light systems, and in large-scale electricity generators.

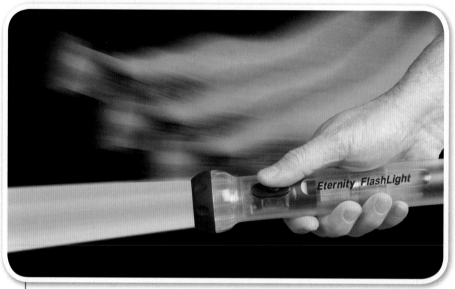

Shaking this torch moves a magnet in and out of a coil to generate an electric current.

The technician is constructing this generator. Wires are wound around the outside. A turbine will make magnets rotate in the centre, inducing an electric current in the wires.

front light

rear light

moving magnet

generator

This bicycle has a small generator that uses the movement of the wheel to produce a current. The generator is connected to a rechargeable battery that supplies its lights.

Questions

3 In a shake torch, how will the current change as the torch is shaken more vigorously? Explain why.

4 Suggest a situation where a wind-up radio would be more convenient than an ordinary radio with batteries.

5 Explain why the bicycle light system shown in the picture needs a rechargeable battery.

Human power

Can we be self-sufficient in energy? Instead of using mains electricity from power stations, could we power all our appliances ourselves?

Pedal power station

In 2009, the BBC television programme 'Bang Goes the Theory' set up a human power station. To generate electricity, 70 cyclists pedalled bicycles. The electricity was used to power appliances in a family house.

As high-power appliances were switched on, the cyclists found it harder to pedal enough to supply the power. The oven and the power shower were the most difficult appliances to run with cycle power. The greater the current supplied by the generator, the harder the cyclists had to pedal.

The cyclists became tired, hot, and sweaty. They needed to eat and drink in order to get enough energy to keep pedalling.

Generating electricity is never 100% efficient. Only some of the energy stored in the cyclists' bodies from the food they had eaten was used to produce electricity. Quite a lot was carried away as heat.

Questions

6 Sketch a Sankey diagram for a cyclist in the human power station. The input is the energy stored in their body from the food they have eaten, and the useful output is the energy carried by the electricity. On your diagram label the wasted energy.

7 A fit cyclist can produce an output power of 200 W.
   a If they can keep this up for 24 hours non-stop, what is their energy output in kWh?
   b If you need 125 kWh per day, how many cyclists would you need to be your 'slaves'?

Each bicycle in the human power station was fixed in place with the back wheel connected to a small generator.

Find out about

- how fossil fuels and biomass fuels can be used to produce electricity
- the sequence of events inside a power station
- why power stations are always less than 100% efficient

Who decides?

Electricity is a secondary energy source. Energy companies, operating under government regulation, generate and distribute it.

Energy companies also make decisions on your behalf. When you boil a kettle, the electricity may have come from any type of primary source.

Burning fuel

In a fossil-fuel power station, coal, gas, or oil is burned to boil water and make high-pressure steam. Biofuels, such as wood, can be used in the same way. In some places, heat is extracted from underground rocks to heat water; this is geothermal energy. Any power station that works like this is known as a **thermal power station**.

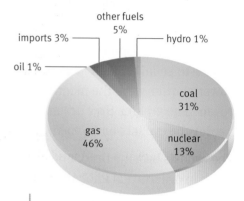

other fuels 5%
imports 3%
hydro 1%
oil 1%
coal 31%
gas 46%
nuclear 13%

In the UK, most electricity nowadays is generated by burning fossil fuels.

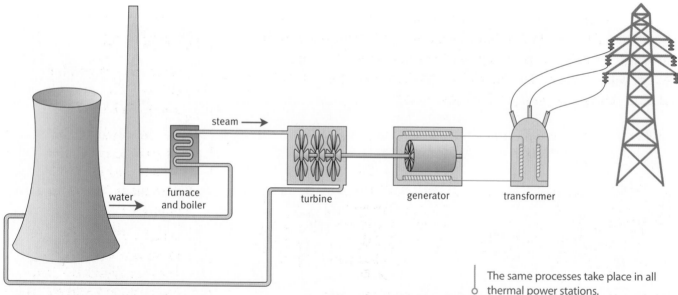

steam →

water   furnace and boiler   turbine   generator   transformer

The same processes take place in all thermal power stations.

Key words

- fossil fuels
- renewable energy sources

Thermal power station

In a thermal power station, high-pressure steam passes through a **turbine**. The turbine rotates a generator to produce electricity. In UK power stations the rate of turning is set at 50 cycles per second.

After passing through the turbines, the steam condenses to water. It can be fed back into the boiler and used again.

This turbine has many small blades that are driven around by the steam.

Regular maintenance keeps the generators running smoothly.

Steam collects in cooling towers where it condenses back to water.

Reducing waste

In any fossil-fuel or biofuel power station, only some of the energy from the burning fuel is transferred electrically. A lot of energy is wasted because it is carried away as heat in steam and exhaust gases.

Burning fuels produce CO_2 and other waste products. Some of the worst pollutants are removed from the exhaust gases before they can escape into the atmosphere.

Using more efficient power stations is one way of reducing the amount of CO_2 produced. Power stations burning natural gas have an extra turbine that is driven by the hot exhaust gases. This makes them the most efficient type of fossil fuel power station. But there are arguments both for and against building more gas-fired power stations.

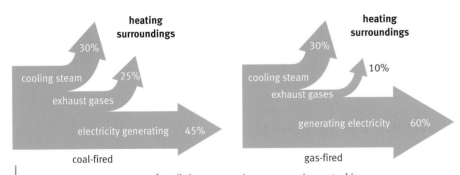

Sankey diagrams account for all the energy. Less energy is wasted in a gas-fired power station.

Weighing the arguments – should we build more gas-fired power stations?

Key words

- ✓ **thermal power station**
- ✓ **turbine**

Questions

1 What percentage of UK electricity is generated using fossil fuels?

2 According to the Sankey diagrams on this page, what is the typical efficiency of a coal-fired power station?

3 Sketch a Sankey diagram to show where the energy goes in a typical oil-fired power station with an efficiency of 38%.

4 What might be the benefits and drawbacks of using biomass fuel in power stations? Draw a balance diagram to summarise your ideas.

There are ten nuclear power stations operating in the UK, some of these are coming towards the end of their working lives. In 2009 the government proposed that another ten should be built and in operation by 2020. They said that these would be essential to meet CO_2 emission targets. Is more nuclear power the right choice?

Nuclear power stations

Nuclear power stations use solid fuel that contains uranium. In a **nuclear reactor**, uranium atoms split into lighter atoms, releasing energy, so the fuel becomes very hot. The hot fuel boils water to make steam that drives turbines.

As the uranium atoms split, the fuel gradually becomes solid nuclear waste.

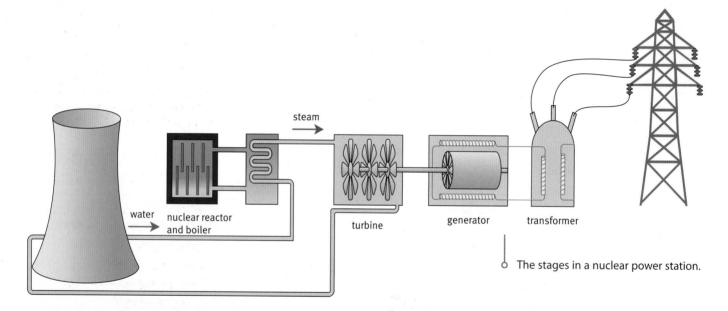

water → nuclear reactor and boiler

steam →

turbine

generator

transformer

The stages in a nuclear power station.

Hot waste from nuclear fuels is stored under water.

Nuclear fuel and waste

Nuclear fuel and nuclear waste are **radioactive**. They give out ionising radiation. Some nuclear waste will be radioactive for thousands of years. Waste from nuclear fuel is very hot at first, so it is stored under water until it cools. When cool, waste is mixed with concrete and stored to make sure it does not contaminate the atmosphere, water supply, or soil.

Nuclear sites are monitored to make sure that workers and the public are not put in danger from radioactive material. Radioactive **contamination** occurs when radioactive material lands on or gets inside something. Exposure to ionising radiation is called **irradiation**. Limits for irradiation are set by law.

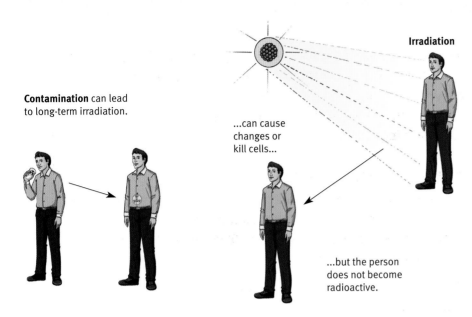

Irradiation

Contamination can lead to long-term irradiation.

...can cause changes or kill cells...

...but the person does not become radioactive.

Key words

- ✓ **nuclear reactor**
- ✓ **radioactive**
- ✓ **irradiation**
- ✓ **contamination**

Uranium mines contain enough fuel for hundreds of years.

Benefits and risks

Nuclear fuel yields far more energy than the same amount of fossil fuel. 1 g of uranium fuel can provide as much energy as 8 kg of fossil fuel.

A nuclear power station produces much less waste than a fossil fuel power station. A nuclear reactor does not burn fuel, so no CO_2 is produced.

Suppose there is an accident and fuel leaks out?

We can't see radiation, how can we judge the risk?

There are no uranium mines in Britain. I am worried about relying on imports.

Why should we be exposed to risks so that everyone can have cheap electricity?

What if nuclear waste falls into the hands of terrorists?

They can't store waste at Sellafield, with sea levels rising. It's on the coast!

Many people are concerned about nuclear power stations.

Questions

1 Why might drinking water, contaminated by radioactive waste, be more dangerous than being irradiated by the glass of water?

2 Do you think that the UK government should build more nuclear power stations? Write an entry for a blog to persuade other people to agree with you.

A **renewable** energy source is one that can be used without running out. We already use some renewable energy sources in the UK. Should we use more?

Solar power

In the UK, electromagnetic radiation from the Sun provides an average of about 100 W of **solar power** per square metre of ground. Solar **thermal** panels use the Sun's radiation to heat water or buildings directly. Covering *all* south-facing roofs in the UK with thermal solar panels could provide about 13 kWh per person per day. A different kind of solar panel uses the Sun's radiation to generate a voltage; these are called **photovoltaic** (PV) panels.

Hydroelectric power

Water heated by the Sun evaporates, and then falls as rain. Rain falling on high ground can be stored behind a dam and used to turn turbines in a **hydroelectric** power station as it flows downhill. The UK gets about 0.2 kWh per person per day from hydroelectric power. If more schemes were built this could rise to 1.5 kWh.

In the UK, placing PV panels on all south-facing roofs could provide 5 kWh per person per day

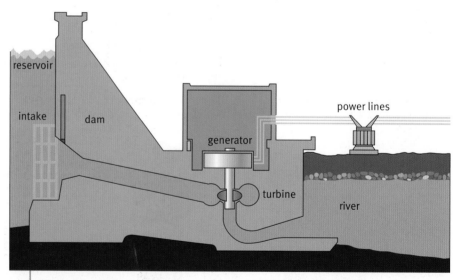

Water from the reservoir turns turbines, which turn the generator.

Wind power

Wind energy can be used to to turn a turbine, which drives an electricity generator. In the UK, a land-based collection of wind turbines (a **wind farm**) has an average output of 2 W per square metre. Wind farms covering the windiest 10% of the country could provide about 20 kWh per person per day. Wind farms built all around the coast of the UK could provide a further 48 kWh but there are major costs and engineering challenges when building at sea.

The Nant y Moch dam is part of a hydroelectric scheme in Wales. The power output from this scheme is 55 MW.

Power from waves and tides

The pull of gravity between the Earth and Moon affects the oceans. As the Moon orbits the Earth, and the Earth rotates, tides rise and fall. Ocean currents are driven by heating from the Sun and by Earth's rotation. Wind, currents, and tides combine to produce waves.

Water movement due to tides and waves can drive turbines. As the UK is surrounded by sea, **tidal power** could provide up to 11 kWh per person per day. There is a tidal energy convertor in Strangford Lough and the government has proposed building a tidal power station in the Severn Estuary.

Biofuels

Biofuels are renewable because they can be replaced quickly. Some biofuels could replace petroleum fuels for transport and all could be burned in thermal power stations. At best, biofuels could provide a total of 7 kWh per person per day in the UK.

Winds are driven by temperature differences in the atmosphere and by the Earth's rotation. The Whitelee wind farm near Glasgow will cover 55 km².

Miscanthus grass is grown for fuel. In the UK each square metre of crop yields about 0.2 W electricity.

In a geothermal power station energy from hot rocks is used to produce steam to drive turbines.

A Pelamis generator uses **wave power** to produce electricity. Pelamis machines along 500 km of Britain's Atlantic coast could produce 4 kWh per person per day.

Key words

- ✓ **renewable**
- ✓ **solar power**
- ✓ **thermal**
- ✓ **hydroelectric**
- ✓ **wind farm**
- ✓ **wave power**
- ✓ **tidal power**

Questions

1 A student says 'All our energy comes from the Sun.' Explain how this is true for the renewable energy sources mentioned on these pages.

2 List the drawbacks of each renewable resource.

3 What would be the maximum total energy in kWh per person per day that we could get from renewable sources in the UK? Suggest reasons why, in practice, the amount is likely to be much less.

Find out about

- ✔ why we need a National Grid
- ✔ why the National Grid uses very high voltages
- ✔ how transformers are used to alter the voltage of an electricity supply

Power stations are built close to their energy source or where there is plenty of cooling water. But that is not always where the electricity is needed. There are energy issues involved in the distribution of electricity, as well as in its generation.

National Grid

All the power stations in the UK are connected to the **National Grid**, which is used to distribute electricity to all the places where we want to use it. It does this by means of a network of long wires and transformers. The power sockets in your home are connected to the Grid. You don't just depend on the nearest power station.

When an electric current flows in a wire, it causes heating. Even if the heating is only slight, it means that energy is wasted rather than getting to the user. The National Grid covers large distances so losses due to heating can be significant.

High voltage

In the UK the domestic supply voltage is 230 V. This is high enough to give a fatal electric shock, but the voltage used over most of the National Grid is very much higher.

To distribute electric power at 230 V, very large currents would be needed. These currents would cause a lot of heating in the cables and so much energy to be wasted.

It is more efficient to use a high voltage to distribute electricity. The higher the voltage, the smaller the current needed for the same power output to the user. With a smaller current, less energy is lost due to heating in the wires.

Transformers

The voltage of an a.c. electricity supply can be altered using a **transformer**.

The high-voltage wires of the Grid are supported by tall pylons to keep them out of reach.

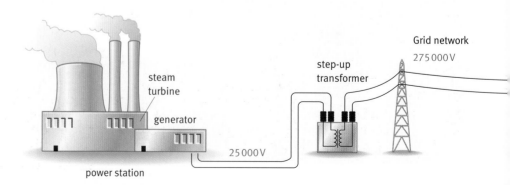

In a step-up transformer the output voltage is higher than the input voltage and the current is reduced. The National Grid uses step-up transformers to increase the voltage for transmission of electricity. A step-down transformer in the local substation reduces the voltage and the current is increased.

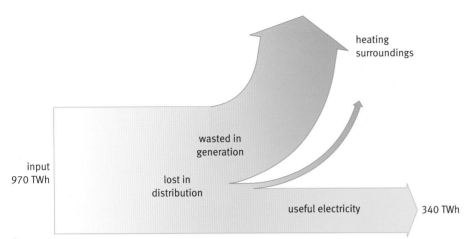

In 2008 the UK used about 970 TWh of energy to supply 340 TWh of electricity. 40 TWh was lost in the distribution process. 1 TWh is one million MWh.

This transformer is about to be installed in a substation.

Questions

1 The UK National Grid is connected to the French grid. Suggest a reason for this.

2 Electricity power lines can be buried under the ground or carried by pylons. Suggest an advantage and a disadvantage of each method.

3 Use the Sankey diagram to calculate the overall efficiency of electricity generation and distribution in the UK.

4 Someone has written to their local paper complaining that large pylons and transformer substations are 'an ugly blot on the landscape' and asking why houses can't just be connected to the nearest power station with 'normal cables'. Write a letter explaining why the pylons and substations are needed.

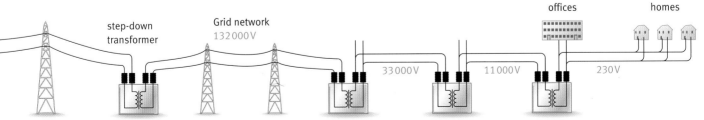

The energy debate

Nuclear

Find out about

- ✓ arguments for and against using various energy sources
- ✓ how to make your views known

YES

Nuclear power is the only energy source that can meet a substantial electricity demand. It releases no damaging carbon dioxide.

The best way to use world stockpiles of uranium and plutonium is as fuel in civilian reactors to generate electricity. Otherwise they remain available for making nuclear weapons.

UK nuclear power stations use tried-and-tested technology. Safety systems meet high standards. Waste disposal is a problem that can be solved.

NO

Nuclear power stations may release little CO_2 while operating. But large amounts of CO_2 are released during construction and decommissioning. Most importantly, they produce radioactive waste.

A new generation of reactors would take about a decade to build and cost roughly £2 billion each. No insurance company will cover their risks, during operation or decommissioning. The public will have to pay if anything goes wrong.

Renewable energy sources

NO

Renewable energy is unreliable. Winds don't always blow. The Sun doesn't always shine.

Renewable sources would not provide enough energy for this country.

No wind farm should be built where people live and work. Each wind turbine is a huge noisy machine.

Wave and tidal generators interfere with wildlife. Large hydroelectric schemes damage the countryside.

The main renewable energy sources are not in the same place as existing power stations. We would have to extend the National Grid with more power lines across the countryside.

YES

The UK should exploit its own energy sources and not rely on imports. Recent studies suggest that renewable energy sources could provide the UK with a reliable supply of electricity.

What we need is a full range of generators – very big to very small – at sites all around the country. A decentralised power system would be based on microgeneration. Installing wind generators and solar cells on the rooftops of many offices and homes will be relatively cheap and easy.

A life cycle assessment shows that power from the Sun and winds releases little CO_2.

Use less energy – for and against

YES

Energy consumption rises year by year. In your lifetime, you are likely to use as much energy as all four of your grandparents put together. Every energy saving you can make will help.

The government can help by:

- requiring new buildings to use less energy for heating and lighting
- providing grants to help householders install domestic combined heat and power systems
- ensuring that new appliances are energy efficient
- taxing fossil fuels more highly.

NO

It's all very well to dream of using less energy. But energy makes the world go round. It's essential to education, business, and pleasure. And everyone has a right to a good standard of living at home.

There isn't a simple answer to the energy question. There will have to be a mix of energy sources, as there is now. No single source can meet all our needs. And that leaves us with more questions:

- Who should make the decisions about which energy sources are in the mix – the energy companies, the government, or scientists?
- Who should have the last word when a few local people object to a new power station that will meet the needs of many more people?
- How do you weigh the benefits of 'clean' energy from wind turbines against the change in view across the hills?

Key word
✓ **decommission**

Questions

1 What does 'a sustainable supply of energy' mean?

2 The cost of decommissioning contributes to the price of electricity. It is much larger for nuclear power stations than for stations burning fossil fuels. Explain why.

3 Look at the arguments for and against each energy source.
   a Draw balance diagrams for each option, listing statements on each side.
   b Distinguish statements of fact from opinion, by putting a tick next to facts.
   c Use the information in this book to add further statements to your diagrams.

4 Write a letter to your Member of Parliament expressing your views about future power stations. Use your answer to question 3 to make your letter persuasive and show that you have considered the issues.

The dismantling of Berkeley nuclear power station in Gloucestershire. Energy costs take in the whole life of a power station, from start to finish.

Science
Explanations

In this module you will learn about the uses of energy from fossil fuels and nuclear power and explore renewable sources of energy. You will also learn how electricity is generated and distributed.

You should know:

- that the demand for energy is continually increasing, and that this raises issues about the availability of sources and the environmental effect of using them
- the main primary energy sources
- why electricity is called a secondary energy source and why it is convenient to use
- which renewable energy sources are used for generating electricity
- that burning carbon fuels in power stations produces carbon dioxide
- that power is the rate at which energy is transferred
- that electrical energy transferred = power × time
- that electric power = voltage × current
- that joules and kilowatt-hours are both units of energy
- how to interpret and construct Sankey diagrams
- that efficiency of electrical appliances and power stations can be calculated using the equation:

$$\text{efficiency} = \frac{\text{energy usefully transferred}}{\text{total energy supplied}}$$

- that mains electricity is produced by generators, which contain coils of wire and spinning magnets
- that thermal power stations use a primary energy source to heat water to drive a turbine and generator, but that many renewable sources of energy drive a turbine directly
- how to label a block diagram showing the main parts of power stations
- that nuclear power stations produce radioactive waste, which emits ionising radiation
- the distinction between contamination and irradiation by a radioactive material
- how electricity is distributed through the National Grid at high voltage, although the main supply voltage to our homes is 230 V
- how to evaluate energy sources, using data where appropriate, in terms of:
  - where they are used (home, work place, or nationally)
  - factors that affect the choice of the source (the environment, economics, waste products produced)
  - the advantages and disadvantages of different non-renewable and renewable power stations (fossil fuel, nuclear, biomass, solar, wind, and water).

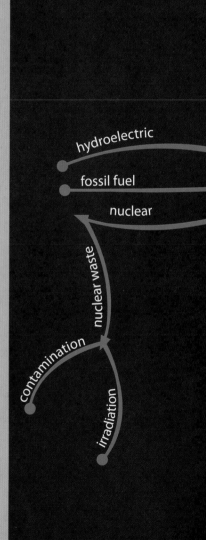

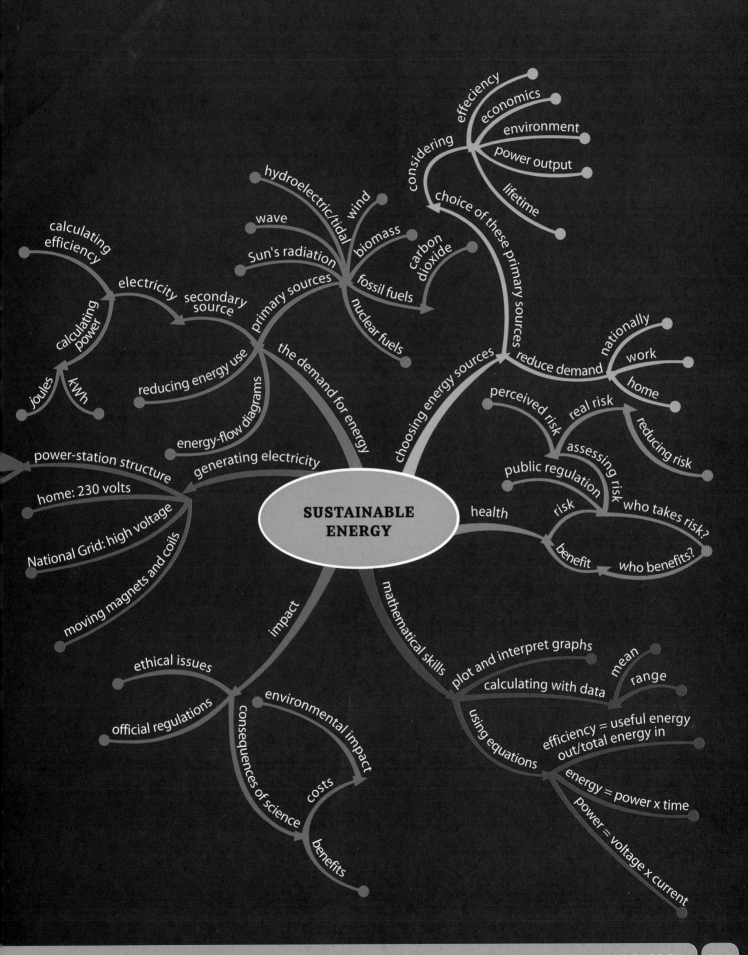

SUSTAINABLE ENERGY

considering
- efficiency
- economics
- environment
- power output
- lifetime

choice of these primary sources

primary sources
- hydroelectric/tidal
- wind
- wave
- biomass
- Sun's radiation
- fossil fuels → carbon dioxide
- nuclear fuels

secondary source
- electricity
  - calculating efficiency
  - calculating power
    - joules
    - kWh

reducing energy use

the demand for energy

energy-flow diagrams

generating electricity

power-station structure
- home: 230 volts
- National Grid: high voltage
- moving magnets and coils

choosing energy sources

reduce demand
- nationally
- work
- home
- reducing risk

perceived risk

real risk

assessing risk

public regulation

who takes risk?

health
- risk
- benefit → who benefits?

impact
- ethical issues
- official regulations
- consequences of science
  - costs
  - benefits
- environmental impact

mathematical skills
- plot and interpret graphs
- calculating with data
  - mean
  - range
- using equations
  - efficiency = useful energy out/total energy in
  - energy = power x time
  - power = voltage x current

Ideas about Science

In addition to developing an understanding of the use and generation of electricity, it is important to assess the risks and benefits associated with the chosen methods of energy use, and to appreciate the issues involved in making decisions about the use of science and technology.

Everything we do carries some risk, and new technologies often introduce new risks. It is important to assess the chance of a particular outcome happening, and the consequences if it did, because people often perceive a risk as being different from the actual risk: sometimes less, and sometimes more. A particular situation that introduces risk will often also introduce benefits, which must be weighed up against that risk.

You should be able to:
- identify risks arising from scientific or technological advances
- suggest ways of reducing a given risk
- interpret and assess risk presented in different ways
- distinguish between real risk and perceived risk
- suggest reasons for given examples of differences between perceived and actual risk
- discuss how risk should be regulated by governments and other public bodies and explain why it may be controversial.

Science-based technology provides people with many things they value. However, some applications of science can have undesirable effects on the quality of life and on the environment. Benefits need to be weighed against costs.

In the context of the sustainable energy, you should be able to:
- identify the groups affected, and the main benefits and costs of a course of action for each group
- suggest reasons why different decisions on the same issue might be taken in different social and economic contexts
- identify examples of unintended impacts of human activity on the environment
- explain the idea of sustainability, and use it to compare the sustainability of different processes
- discuss the official regulation of the application of scientific knowledge
- in cases where an ethical issue is involved, say clearly what the issue is and summarise different views that may be held
- understand ethical arguments based on 'the greater good of the greater number' and that certain things are right or wrong whatever the circumstances.

Review Questions

1

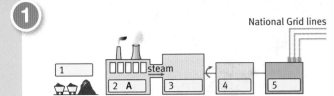

National Grid lines

a Copy and label the diagram of a coal-fired power station.
Put the letters **A**, **B**, **C**, **D**, and **E** in the correct boxes on the diagram. One has been done for you.

A	furnace
B	transformer
C	fuel
D	turbine
E	generator

b Power stations use a carbon-based fuel. Which greenhouse gas will definitely be produced when the fuel is burnt?

c Coal is a non-renewable energy source. Which two of the following are **renewable** energy sources that are used to generate electricity?

natural gas   **nuclear fuel**
wind power   **oil**   **wave power**

2 The diagram shows the efficiency of a modern power station.

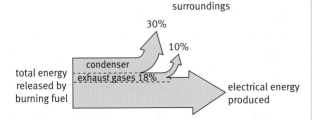

a Use the diagram to calculate the efficiency of the power station in producing electrical energy.

b One way to waste less energy is to use the heat energy from the condenser to heat homes and businesses near the power station. Assuming **half** of the heat from the condenser can be used in this way, what is the efficiency of the power station in providing useful energy output?

3 An electric heater draws a current of 10 A from a 230 V power supply.

a Calculate the input power, in watts and in kilowatts, to the heater.

b Calculate the cost of using the heater for five hours, if one kilowatt-hour of electrical energy costs 8 p.

4 Many wind farms are being planned to generate electricity for Britain.
The pie chart below shows the various costs of setting up and operating a wind farms.

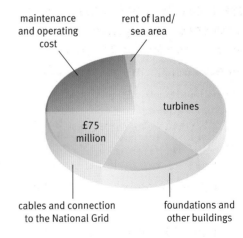

a Which one of the following is the best estimate of the cost of the turbines?

£40 million   **£75 million**
£150 million   **£200 million**

b Which one of the factors costs £90 million?

Glossary

absorb (radiation) The radiation that hits an object and is not reflected, or transmitted through it, is absorbed (for example, black paper absorbs light). Its energy makes the object get a little hotter.

absorption spectrum (of a star) Consists of dark lines superimposed on a continuous spectrum. It is created when light from the star passes through a cooler gas that absorbs photons of particular energies.

accumulate To collect together and increase in quantity.

accuracy How close a quantitative result is to the true or 'actual' value.

actual risk Risk calculated from reliable data.

adaptation A feature that helps an organism survive in its environment.

ADH A hormone that makes kidney tubules more permeable to water, causing greater re-absorption of water.

aerial A wire, or arrangement of wires, that emits radio waves when there is an alternating current in it, and in which an alternating current is induced by passing radio waves. So it acts as a source or a receiver of radio waves.

alcohol The intoxicating chemical in wine, beer, and spirits. Causes changes in behaviour and may create long-term addiction.

Alkali Acts The Acts of Parliament passed in the UK in order to control levels of pollution. They led to the formation of an Alkali Inspectorate, which checked that at least 95% of acid fumes were removed from the chimneys of chemical factories.

alkali A compound that dissolves in water to give a solution with a pH higher than 7. An alkali can be neutralised by an acid to form a salt. Solutions of alkalis contain hydroxide ions.

allele Different versions of the same gene.

allergic People with an allergy suffer symptoms when they eat some foods that most people find harmless. Symptoms can include itchy skin, shortness of breath, and an upset stomach.

ammeter A meter that measures the size of an electric current in a circuit.

ampere (amp, for short) The unit of electric current.

amplifier A device for increasing the amplitude of an electrical signal. Used in radios and other audio equipment.

amplitude For a mechanical wave, the maximum distance that each point on the medium moves from its normal position as the wave passes. For an electromagnetic wave, the maximum value of the varying electric field (or magnetic field).

analogue signal Signals used in communications in which the amplitude can vary continuously.

antibiotic resistant Microorganisms that are not killed by antibiotics.

antibiotics Drugs that kill or stop the growth of bacteria and fungi.

antibodies A group of proteins made by white blood cells to fight dangerous microorganisms. A different antibody is needed to fight each different type of microorganism. Antibodies bind to the surface of the microorganism, which triggers other white blood cells to digest them.

antigens The proteins on the surface of a cell. A cell's antigens are unique markers.

artery A blood vessel that carries blood away from the heart.

asexual reproduction When a new individual is produced from just one parent.

assumption A piece of information that is taken for granted without sufficient evidence to be certain.

asteroid A dwarf rocky planet, generally orbiting the Sun between the orbits of Mars and Jupiter.

atmosphere The layer of gases that surrounds the Earth.

atom The smallest particle of an element. The atoms of each element are the same as each other and are different from the atoms of other elements.

best estimate When measuring a variable, the value in which you have most confidence.

big bang An explosion of a single mass of material. This is currently the accepted scientific explanation for the start of the Universe.

biodegradable Materials that are broken down in the environment by microorganisms. Most synthetic polymers are *not* biodegradable.

biodiversity The great variety of living things, both within a species and between different species.

biofuel A renewable fuel that uses biological material, such as recently living plant materials and animal waste.

bleach A chemical that can destroy unwanted colours. Bleaches also kill bacteria. A common bleach is a solution of chlorine in sodium hydroxide.

blind trial A clinical trial in which the patient does not know whether they are taking the new drug, but their doctor does.

blood pressure The pressure exerted by blood pushing on the walls of a blood vessel.

blood transfusion Transfer of blood from one person to another.

branched chain A chain of carbon atoms with short side branches.

brine A solution of sodium chloride (salt) in water. Brine is produced by solution mining of underground salt deposits.

capillary The smallest blood vessel. Its walls are only one cell thick and allow substances to diffuse between the blood and the cells.

carbon cycle The cycling of the element carbon in the environment between the atmosphere, biosphere, hydrosphere, and lithosphere. The element exists in different compounds in these spheres. In the atmosphere it is mainly present as carbon dioxide.

carrier A steady stream of radio waves produced by an RF oscillator in a radio to carry information.

carrier Someone who has the recessive allele for a characteristic or disease but who does not have the characteristic or disease itself.

catalyst A chemical that speeds up a chemical reaction but is not used up in the process.

catalytic converter A device fitted to a vehicle exhaust that changes the waste gases into less harmful ones.

cause When there is evidence that changes in a factor produce a particular outcome, then the factor is said to cause the outcome. For example, increases in the pollen count cause increases in the incidence of hayfever.

centrifuge A piece of equipment used to separate a mixture of liquids and solids by spinning the mixture very fast.

ceramic Solid material such as pottery, glass, cement, and brick.

chemical change/reaction A change that forms a new chemical.

chemical equation A summary of a chemical reaction showing the reactants and products with their physical states (see balanced chemical equation).

chemical formula A way of describing a chemical that uses symbols for atoms. It gives information about the number of different types of atom in the chemical.

chemical synthesis Making a new chemical by joining together simpler chemicals.

chlorination The process of adding chlorine to water to kill microorganisms, so that it is safe to drink.

chlorine A greenish toxic gas, used to bleach paper and textiles, and to treat water.

chlorofluorocarbons (CFCs) Liquids that used to be used in refrigerators and aerosols. Their vapour damages the ozone layer.

chromosome Long, thin, threadlike structure in the nucleus of a cell made from a molecule of DNA. Chromosomes carry the genes.

classification Putting living things into groups based on their shared characteristics.

climate Average weather in a region over many years.

clinical trial When a new drug is tested on humans to find out whether it is safe and whether it works.

clone A new cell or individual made by asexual reproduction. A clone has the same genes as its parent.

coding (in communications) Converting information from one form to another, for example, changing an analogue signal into a digital one.

combustion The process of burning a substance that reacts with oxygen to produce heat and light.

comet A rocky lump, held together by frozen gases and water, that orbits the Sun.

competition Different organisms that require the same resource, such as water, food, light, or space, must compete for the resource.

compression A material is in compression when forces are trying to push it together and make it smaller.

computer model A computer uses data and equations to study events that have happened or to predict what might happen.

concentration The quantity of a chemical dissolved in a stated volume of solution. Concentrations can be measured in grams per litre.

condensed The change of state from a gas to a liquid, for example, water vapour in the air condenses to form rain.

conservation of atoms All the atoms present at the beginning of a chemical reaction are still there at the end. No new atoms are created and no atoms are destroyed during a chemical reaction.

conservation of energy The principle that the total amount of energy at the end of any process is always equal to the total amount of energy at the beginning – though it may now be stored in different ways and in different places.

conservation of mass The total mass of chemicals is the same at the end of a reaction as at the beginning. No atoms are created or destroyed and so no mass is gained or lost.

consumer An organism that eats others in a food chain. This is all the organisms in a food chain except the producer(s).

continental drift A theory that describes the extremely slow movements of the continents across the Earth.

control In a clinical trial, the control group is people taking the currently used drug. The effects of the new drug can then be compared to this group.

convection The movement that occurs when hot material rises and cooler material sinks.

core The Earth's core is made mostly from iron, solid at the centre and liquid above.

coronary artery The artery that supplies blood carrying oxygen and glucose directly to the muscle cells of the heart.

correlation A link between two things. For example, if an outcome happens when a factor is present, but not when it is absent, or if an outcome increases or decreases when a factor increases. For example, when pollen count increases hayfever cases also increase.

cross-link A link or bond joining polymer chains together.

crude oil A dark, oily liquid found in the Earth, which is a mixture of hydrocarbons.

crust A rocky layer at the surface of the Earth, 10–40 km deep.

crystalline polymer A polymer with molecules lined up in a regular way as in a crystal.

crystallise To form crystals, for example, by evaporating the water from a solution of a salt.

cystic fibrosis An inherited disorder. The disorder is caused by recessive alleles.

decoding In communications, converting information back into its original form, for example, changing a digital signal back into an analogue one.

decommissioning Taking a power station out of service at the end of its lifetime, dismantling it, and disposing of the waste safely.

decomposer bacteria Microorganisms that break down the organic compounds in dead plants and animals and waste.

decomposition The process of breaking down dead plants and animals and waste by microorganisms.

deforestation Cutting down trees from an area of land.

denitrification The removal of nitrogen from soil. Bacteria break down nitrates in the soil, converting them back to nitrogen.

denitrifying bacteria Bacteria that break down nitrates in the soil, releasing nitrogen into the air.

density A dense material is heavy for its size. Density is mass divided by volume.

detector Any device or instrument that shows the presence of radiation by absorbing it.

detritivore An organism that feeds on dead organisms and waste. Woodlice, earthworms, and millipedes are examples of detritivores.

digest To break down larger, insoluble molecules into small, soluble molecules.

digital code A string of 0s and 1s that can be used to represent an analogue signal, and from which that signal can be reconstructed.

digital signal Signals used in communications in which the amplitude can take only one of two values, corresponding to the digits 0 and 1.

disease A condition that impairs normal functioning of an organism's body, usually associated with particular signs and symptoms. It may be caused by an infection or by the dysfunction of internal organs.

dissolve Some chemicals dissolve in liquids (solvents). Salt and sugar, for example, dissolve in water.

DNA (deoxyribonucleic acid) The chemical that makes up chromosomes. DNA carries the genetic code, which controls how an organism develops.

dominant Describes an allele that will show up in an organism even if a different allele of the gene is present. You only need to have one copy of a dominant allele to have the feature it produces.

double-blind trial A clinical trial in which neither the doctor nor the patient knows whether the patient is taking the new drug.

durable A material is durable if it lasts a long time in use. It does not wear out.

duration How long something happens for. For example, the length of time someone is exposed to radiation.

dwarf planet A round, planetlike object with a similar orbit to the eight planets but too small to clear its orbit of other small objects.

earthquake An event in which rocks break to allow tectonic plate movement, causing the ground to shake.

economic context How money changes hands between businesses, government, and individuals.

Ecstasy A recreational drug that increases the concentration of serotonin at the synapses in the brain, giving pleasurable feelings. Long-term effects may include destruction of the synapses.

effector The part of a control system that brings about a change to the system.

efficiency The percentage of the energy supplied to a device that is transferred to the desired place, or in the desired way.

electric charge A fundamental property of matter. Electrons and protons are charged particles. Objects become charged when electrons are transferred to or from them, for example, by rubbing.

electric current A flow of charge around an electric circuit.

electrolysis Splitting up a chemical into its elements by passing an electric current through it.

electromagnetic induction The name of the process in which a potential difference (and hence often an electric current) is generated in a wire, when it is in a changing magnetic field.

electromagnetic spectrum The 'family' of electromagnetic waves of different frequencies and wavelengths.

electromagnetic wave A wave consisting of vibrating electric and magnetic fields, which can travel in a vacuum. Visible light is one example.

electron A tiny, negatively charged particle, which is part of an atom. Electrons are found outside the nucleus. Electrons have negligible mass and one unit of charge.

embryo The earliest stage of development for an animal or plant. In humans the embryo stage lasts for the first two months.

emission Something given out by something else, for example, the emission of carbon dioxide from combustion engines.

emit Give out (radiation).

endangered A species that is at risk of becoming extinct.

energy cost The amount of energy used to produce something or to do something.

environment Everything that surrounds you. This is factors like the air and water, as well as other living things.

enzyme A protein that catalyses (speeds up) chemical reactions in living things.

epidemiological study A scientific study that examines the causes, spread, and control of a disease in a human population.

erosion The movement of solids at the Earth's surface (for example, soil, mud, and rock) caused by wind, water, ice, and gravity, and living organisms.

ethics A set of principles that may show how to behave in a situation.

evaporate The change of state from a liquid to a gas.

evolution The process by which species gradually change over time. Evolution can produce new species.

excretion The removal of waste products of chemical reactions from cells.

extinct A species is extinct when all the members of the species have died out.

extruded A plastic is shaped by being forced through a mould.

factor A variable that changes and may affect something else.

false negative A wrong test result. The test result says that a person does not have a medical condition but this is incorrect.

false positive   A wrong test result. The test result says that a person has a medical condition but this is incorrect.

fertile   An organism that can produce offspring.

fibres   Long thin threads that make up materials such as wool and polyester. Most fibres used for textiles consist of natural or synthetic polymers.

filter   To separate a solid from a liquid by passing it through a filter paper.

flavouring   Mixtures of chemicals that give food, sweets, toothpaste, and other products their flavours.

flexible   A flexible material bends easily without breaking.

food chain   In the food industry this covers all the stages from where food grows, through harvesting, processing, preservation, and cooking to being eaten.

food web   A series of linked food chains showing the feeding relationships in a habitat – 'what eats what'.

formula (chemical)   A way of describing a chemical that uses symbols for atoms. A formula gives information about the numbers of different types of atom in the chemical. The formula of sulfuric acid, for example, is H_2SO_4.

fossil   The stony remains of an animal or plant that lived millions of years ago, or an imprint it has made (for example, a footprint) in a surface.

fossil fuel   Natural gas, oil, or coal.

fraction   A mixture of hydrocarbons with similar boiling points that have been separated from crude oil by fractional distillation.

fractional distillation   The process of separating crude oil into groups of molecules with similar boiling points called fractions.

fuel rod   A container for nuclear fuel, which enables fuel to be inserted into, and removed from, a nuclear reactor while it is operating.

functional protein   Proteins that take part in chemical reactions, for example, enzymes.

fungi   A group of living things, including some microorganisms, that cannot make their own food.

galaxy   A collection of thousands of millions of stars held together by gravity.

gamma radiation (gamma rays)   The most penetrating type of ionising radiation, produced by the nucleus of an atom in radioactive decay. The most energetic part of the electromagnetic spectrum.

gene   A section of DNA giving instructions for a cell about how to make one kind of protein.

generator   A device used to produce electricity by spinning a magnet near a coil of wire (or a coil near a magnet).

genetic screening   Testing a population for a particular allele.

genetic study   A scientific study of the genes carried by people in a population to look for alleles that increase the risk of disease.

genetic test   A test to find whether a person has a particular DNA sequence or allele.

genetic variation   The differences between individuals caused by differences in their genes. Gametes show genetic variation – they all have different genes.

genotype   A description of the genes an organism has.

geothermal power station   A power station that uses energy from hot underground rocks to heat water to produce steam to drive turbines.

globular cluster   A cluster of hundreds of thousands of old stars.

grain   A relatively small particle of a substance, for example, grains of sand.

greenhouse effect   The atmosphere absorbs infrared radiation from the Earth's surface and radiates some of it back to the surface, making it warmer than it would otherwise be.

greenhouse gas   A gas that contributes to the greenhouse effect, including carbon dioxide, methane, and water vapour.

habitat   The place where an organism lives.

hard   A material that is difficult to dent or scratch.

heart disease   A disease where the coronary arteries become increasingly blocked with fatty deposits, restricting the blood flow to the heart muscle. The risk of this is increased by a high fat diet, smoking, and drinking excess alcohol.

heterozygous   An individual with two different alleles for a particular gene.

homeostasis   Keeping a steady state inside your body.

homozygous   An individual with both alleles of a particular gene the same.

human trial   The stage of the trial process for a new drug where the drug is taken by healthy volunteers to see if it is safe, and then by sick volunteers to check that it works.

Huntington's disease   An inherited disease of the nervous system. The symptoms do not show up until middle age.

hydrocarbon   A compound of hydrogen and carbon only. Ethane, C_2H_6, is a hydrocarbon.

hydroelectric power station   A power station that uses water stored behind a dam to drive turbines to generate electricity.

hydrogen chloride gas   An acid gas that is toxic and corrosive, and is produced by the Leblanc process.

hydrogen sulfide gas   A poisonous gas that smells of rotten eggs.

immune   Able to react to an infection quickly, stopping the microorganisms before they can make you ill, usually because you've been exposed to them before.

immune system   A group of organs and tissues in the body that fight infections.

incinerator   A factory for burning rubbish and generating electricity.

indirectly   When something humans do affects another species, but this wasn't the reason for the action. For example, a species habitat is destroyed when land is cleared for farming.

infectious   A disease that can be caught. The microorganism that causes it is passed from one person to another through the air, through water, or by touch.

infertile   An organism that cannot produce offspring.

information (in a computer)   Data stored and processed. It is measured in bytes.

infrared   Electromagnetic waves with a frequency lower than that of visible light, beyond the red end of the visible spectrum.

inherited   A feature that is passed from parents to offspring by their genes.

interdependence   The relationships between different living things that they rely on to survive.

intrinsic brightness (of a star)   A measure of the light that would reach a telescope if a star were at a standard distance from the Earth.

ion   An electrically charged atom, or group of atoms.

ionising radiation   Radiation with photons of sufficient energy to remove electrons from atoms in its path. Ionising radiation, such as ultraviolet, X-rays, and gamma rays, can damage living cells.

irradiation   Being exposed to radiation from an external source.

joule   A unit used to measure energy.

kidneys   Organs in the body that removes waste urea from the blood, and balances water and blood plasma levels. People are usually born with two kidneys.

kilowatt-hour   A unit used to measure energy. It is equivalent to using energy at a rate of 1 kilowatt for 1 hour. Power (kW) × time (hours).

landfill   Disposing of rubbish in holes in the ground.

latitude   The location of a place on Earth, north or south of the equator.

leach   The movement of the plasticisers in a polymer into water, or another liquid, that is flowing past the polymer or is contained by it.

Leblanc process   A process that used chalk (calcium carbonate), salt (sodium chloride) and coal to make the alkali, sodium carbonate. The Leblanc process was highly polluting.

lichen   An organism consisting of a fungus growing with a simple photosynthetic organism called an alga. Lichens grow very slowly are often found growing on walls and roofs.

life cycle assessment   A way of analysing the production, use, and disposal of a material or product to add up the total energy and water used and the effects on the environment.

lifestyle   The way in which people choose to live their lives, for example, what they choose to eat, how much exercise they choose to do, how much stress they experience in their job.

lifestyle disease   a disease that is not caused by microorganisms. They are triggered by other factors, for example, smoking, diet, and lack of exercise.

light pollution   Light created by humans, for example, street lighting, that prevents city dwellers from seeing more than a few bright stars. It also causes problems for astronomers.

light-year   The distance travelled by light in one year.

long-chain molecule   Polymers are long-chain molecules. They consist of long chains of atoms.

longitudinal wave   A wave in which the particles of the medium vibrate in the same direction as the wave is travelling. Sound is an example.

macroscopic Large enough to be seen without the help of a microscope.

magnetic A material that is attracted to a magnet. For example, iron is magnetic.

mantle A thick layer of rock beneath the Earth's crust, which extends about halfway down to the Earth's centre.

match Some studies into diseases compare two groups of people. People in each group are chosen to be as similar as possible (matched) so that the results can be fairly compared.

material The polymers, metals, glasses, and ceramics that we use to make all sorts of objects and structures.

mayfly larvae Mayflies spend most of their lives (up to three years) as larvae (also called mayfly nymphs). They live and feed in aquatic environments. The adult insects live on the wing for a short time, from a few hours to a few days.

mean value A type of average, found by adding up a set of measurements and then dividing by the number of measurements. You can have more confidence in the mean of a set of measurements than in a single measurement.

mechanism A process that explains why a particular factor causes an outcome.

medium (plural media) A material through which a wave travels.

melting point The temperature at which something melts.

memory cell A long-lived white blood cell, which is able to respond very quickly (by producing antibodies to destroy the microorganism) when it meets a microorganism for the second time.

metal The elements on the left side of the periodic table. Metals have characteristic properties: they are shiny when polished and they conduct electricity. Some metals react with acids to give salts and hydrogen. Metals are present as positive ions in salts.

microorganism A living organism that can only be seen through a microscope. They include bacteria, viruses, and fungi.

microwave The radio wave with the highest frequency (shortest wavelength), used for mobile phones and satellite TV.

Milky Way The galaxy in which the Sun and its planets including Earth are located. It is seen from the Earth as an irregular, faintly luminous band across the night sky.

mixture Two or more different chemicals, mixed but not chemically joined together.

modulate To vary the amplitude or frequency of carrier waves so that they carry information.

molecule A group of atoms joined together. Most non-metals consist of molecules. Most compounds of non-metals with other non-metals are also molecular.

monoculture The continuous growing of one type of crop.

monomer A small molecule that can be joined to others like it in long chains to make a polymer.

mountain chain A group of mountains that extend along a line, often hundreds or even thousands of kilometres. Generally caused by the movement of tectonic plates.

mutation A change in the DNA of an organism. It alters a gene and may change the organism's characteristics.

nanometre A unit of length 1 000 000 000 times smaller than a metre. 1 nm = 0.001 μm = 0.000001 mm.

nanoparticle A very tiny particle, whose size can be measured in nanometres.

nanotechnology The control of matter on a tiny (nanometre) scale.

National Grid A network of cables and transformers that connects power stations to the consumers who use the electricity.

natural A material that occurs naturally but may need processing to make it useful, such as silk, cotton, leather, and asbestos.

natural selection When certain individuals are better suited to their environment they are more likely to survive and breed, passing on their features to the next generation.

negative feedback A system where any change results in actions that reverse the original change.

neutralisation A reaction in which an acid reacts with an alkali to form a salt. During neutralisation reactions, the hydrogen ions in the acid solution react with hydroxide ions in the alkaline solution to make water molecules.

nitrogen cycle The continual cycling of nitrogen, which is one of the elements essential for life. By being converted to different chemical forms, nitrogen is able to pass between the atmosphere, lithosphere, hydrosphere, and biosphere.

nitrogen fixation When nitrogen in the air is converted into nitrates in the soil by bacteria.

nitrogen-fixing bacteria Bacteria found in the soil and in swellings (nodules) on the roots of some plants (legumes), such as clover and peas. These bacteria take in nitrogen gas and make nitrates, which plants can absorb and use to make proteins.

noise Unwanted electrical signals that get added on to radio waves during transmission, causing additional modulation. Sometimes called 'interference'.

non-ionising radiation Radiation with photons that do not have enough energy to ionise molecules.

nuclear fission The process in which a nucleus of uranium-235 breaks apart, releasing energy, when it absorbs a neutron.

nuclear fuel In a nuclear reactor, each uranium atom in a fuel rod undergoes fission and releases energy when hit by a neutron.

nuclear fusion The process in which two small nuclei combine to form a larger one, releasing energy. An example is hydrogen combining to form helium. This happens in stars, including the Sun.

nucleus The central core of the atom. It is made up of protons and neutrons.

nucleus (plural nuclei) The central structure in a cell containing genetic material. It controls the function and characteristics of the cell.

observed brightness (of a star) A measure of the light reaching a telescope from a star.

oceanic ridge A line of underwater mountains in an ocean, where new seafloor constantly forms.

open-label trial A clinical drug test in which both the patient and their doctor knows whether the patient is taking the new drug.

optical fibre A thin glass fibre, down which a light beam can travel. The beam is reflected at the sides by total internal reflection, so very little escapes. Used in modern communications, for example, to link computers in a building to a network.

organic matter Material that has come from dead plants and animals.

outcome A variable that changes as a result of something else changing.

outlier A measured result that seems very different from other repeat measurements, or from the value you would expect, which you therefore strongly suspect is wrong.

oxidation A reaction that adds oxygen to a chemical.

ozone layer A thin layer in the atmosphere, about 30 km up, where oxygen is in the form of ozone molecules. The ozone layer absorbs ultraviolet radiation from sunlight.

parallax The apparent shift of an object against a more distant background, as the position of the observer changes. The further away an object is, the less it appears to shift. This can be used to measure how far away an object is, for example, to measure the distance to stars.

parallax angle When observed at an interval of six months, a star will appear to move against the background of much more distant stars. Half of its apparent angular motion is called its parallax angle.

particulate A tiny bit of a solid.

passenger-kilometre A unit used to compare different transport systems to take account of how many passengers are carried. Number of passengers × distance travelled in km.

patent An exclusive right granted for an invention, which may be a product, a process that provides a new way of doing something, or a new technical solution to a problem.

peak frequency The frequency with the greatest intensity.

peer review The process whereby scientists who are experts in their field critically evaluate a scientific paper or idea before and after publication.

persistent organic pollutant (POP) A POP is an organic compound that does not break down in the environment for a very long time. POPs can spread widely around the world and build up in the fatty tissue of humans and animals. They can be harmful to people and the environment.

phenotype A description of the physical characteristics that an organism has (often related to a particular gene).

photon A tiny 'packet' of electromagnetic radiation. All electromagnetic waves are emitted and absorbed as photons. The energy of a photon is proportional to the frequency of the radiation.

photosynthesis A chemical reaction that happens in green plants using the energy in sunlight. The plant takes in water and carbon dioxide, and uses sunlight to convert them to glucose (a nutrient) and oxygen.

photovoltaic (PV) panel   A device that uses the Sun's radiation to generate electricity.

phthalate   A chemical that is used as a plasticiser, added to polymers to make them more flexible.

phytoplankton   Single-celled photosynthetic organisms found in an ocean ecosystem.

pituitary gland   The part of the human brain that coordinates many different functions, for example, release of ADH.

placebo   Occasionally used in clinical trials, this looks like the drug being tested but contains no actual drug.

planet   A very large, spherical object that orbits the Sun, or other star.

plasticiser   A chemical (usually a small molecule) added to a polymer to make it more flexible.

pollutant   A chemical that contaminates the air, water, or soil.

polymer   A material made of very long molecules formed by joining lots of small molecules, called monomers, together.

polymerise   The joining together of lots of small molecules called monomers to form a long-chain molecule called a polymer.

population   A group of animals or plants of the same species living in the same area.

potential difference (p.d.)   The difference in potential energy (for each unit of charge flowing) between any two points in an electric circuit. Also called voltage.

power   In an electric circuit, the rate at which work is done by the battery or power supply on the components in a circuit. Power is equal to current × voltage.

predator   An animal that kills other animals (its prey) for food.

pre-implantation genetic diagnosis (PGD)   This is the technical term for embryo selection. Embryos fertilised outside the body are tested for genetic disorders. Only healthy embryos are put into the mother's uterus.

preservative   A chemical added to food to stop it going bad.

primary energy source   A source of energy not derived from any other energy source, for example, fossil fuels or uranium.

principal frequency   The frequency that is emitted with the highest intensity.

processing centre   The part of a control system that receives and processes information from the receptor, and triggers action by the effectors.

producer   The organism found at the start of a food chain. Producers are autotrophs, able to make their own food.

product   A new chemical formed during a chemical reaction.

properties   The physical or chemical characteristics of a chemical. The properties of a chemical are what make it different from other chemicals.

proportional   Two variables are proportional if there is a constant ratio between them.

protein   Chemicals in living things that are polymers made by joining together amino acids.

pulse rate   The rate at which the heart beats. The pulse is measured by pressing on an artery in the neck, wrist, or groin.

P-wave   A longitudinal seismic wave through the Earth, produced during an earthquake.

radiation   A flow of information and energy from a source. Light and infrared are examples. Radiation spreads out from its source, and may be absorbed or reflected by objects in its path. It may also go (be transmitted) through them.

radio wave   Electromagnetic wave of a much lower frequency than visible light. Radio waves can be made to carry signals and are widely used for communications.

radioactive   Used to describe a material, atom, or element that produces alpha, beta, or gamma radiation.

random   Of no predictable pattern.

range   The difference between the highest and the lowest of a set of measurements.

ray diagram   A way of representing how a lens or telescope affects the light that it gathers, by drawing the rays (which can be thought of as very narrow beams of light) as straight lines.

reactant   A chemical on the left-hand side of an equation. These chemicals react to form the products.

reacting mass   The masses of chemicals that react together, and the masses of products that are formed. Reacting masses are calculated from the balanced symbol equation using relative atomic masses and relative formula masses.

reactive metal   A metal with a strong tendency to react with chemicals such as oxygen, water, and acids. The more reactive a metal, the more strongly it joins with other elements such as oxygen. So reactive metals are hard to extract from their ores.

real difference   The difference between two mean values is real if their ranges do not overlap.

receptor   The part of a control system that detects changes in the system and passes this information to the processing centre.

recessive   An allele that will only show up in an organism when a dominant allele of the gene is not present. You must have two copies of a recessive allele to have the feature it produces.

recycling   A range of methods for making new materials from materials that have already been used.

redshift   When radiation is observed to have longer wavelengths than expected. (Red light has the longest wavelength of visible light.)

reducing agent   A chemical that removes oxygen from another chemical. For example, carbon acts as a reducing agent when it removes oxygen from a metal oxide. The carbon is oxidised to carbon monoxide during this process.

reduction   A reaction that removes oxygen from a chemical.

reference material   A known chemical used in analysis for comparison with unknown chemicals.

regulation   A rule that can be enforced by an authority, for example, the government. The law that says that all vehicles that are three years old and older must have an annual exhaust emission test is a regulation that helps to reduce atmospheric pollution.

renewable energy source   A resource that can be used to generate electricity without being used up, such as the wind, tides, and sunlight.

repeatable   A quality of a measurement that gives the same result when repeated under the same conditions.

reproducible   A quality of a measurement that gives the same result when carried out under different conditions, for example, by different people or using different equipment or methods.

reproduction   The production of offspring through a sexual or asexual process.

reproductive isolation   Two populations are reproductively isolated if they are unable to breed with each other.

respiration   A series of chemical reactions in cells that release energy for the cell to use.

risk   The probability of an outcome that is seen as undesirable, associated with some behaviour or process.

risk factor   A variable linked to an increased risk of disease. Risk factors are linked to disease but may not be the cause of the disease.

rock cycle   Continuing changes in rock material, caused by processes such as erosion, sedimentation, compression, and heating.

rubber   A material that is easily stretched or bent. Natural rubber is a natural polymer obtained from latex, the sap of a rubber tree.

salt   An ionic compound formed when an acid neutralises an alkali or when a metal reacts with a non-metal.

sampling   In the context of physics, measuring the amplitude of an analogue signal many times a second in order to convert it into a digital signal.

Sankey diagram   A flow diagram used to show what happens to energy during a process. The width of the arrows are proportional to the energy flow.

seafloor spreading   The process of forming new ocean floor at oceanic ridges.

secondary energy source   Energy in a form that can be distributed easily but is manufactured by using a raw energy resource such as a fossil fuel or wind. Examples of secondary energy sources are electricity, hot water used in heating systems, and steam.

sedimentary rock   Rock formed from layers of sediment.

seismic wave   A wave produced by the vibrations caused by an earthquake.

selective absorption   Some materials absorb some forms of electromagnetic radiation but not others. For example, glass absorbs infrared but is transparent to visible light.

selective breeding   Choosing parent organisms with certain characteristics and mating them to try to produce offspring that have these characteristics.

sensitivity The ability to detect small changes, for example, radiation or temperature.

sex cells Cells produced by males and females for reproduction – sperm cells and egg cells. Sex cells carry a copy of the parent's genetic information. They join together at fertilisation.

sexual reproduction Reproduction where the sex cells from two individuals fuse together to form an embryo.

shielding Materials used to absorb radiation.

signal Information carried through a communication system, for example, by an electromagnetic wave with variations in its amplitude or frequency, or being rapidly switched on and off.

social context The situation of people's lives.

soft A material that is easy to dent or scratch.

solar power Power supplied by electromagnetic radiation from the Sun.

Solar System The Sun and objects that orbit around it – planets and their moons, comets, and asteroids.

solution Formed when a solid, liquid, or gas dissolves in a solvent.

source An object that produces radiation.

specialised A specialised cell is adapted for a particular job.

species A group of organisms that can breed to produce fertile offspring.

spectrum One example is the continuous band of colours, from violet to red, produced by shining white light through a prism. Passing light from a flame test through a prism produces a line spectrum.

speed of light 300 000 kilometres per second – the speed of all electromagnetic waves in a vacuum.

speed of recession The speed at which a galaxy is moving away from us.

stem cell An unspecialised animal cell that can divide and develop into a specialised cell.

stiff A material that is difficult to bend or stretch.

strong A material that is hard to pull apart or crush.

strong (nuclear) force A fundamental force of nature that acts inside atomic nuclei.

structural Making up the structure (of a cell or organism).

structural protein A protein that is used to build cells.

subsidence The sinking of the ground's surface when it collapses into a hole beneath it.

Sun The star nearest Earth. Fusion of hydrogen in the Sun releases energy, which makes life on Earth possible.

supernova A dying star that explodes violently, producing an extremely bright astronomical object for weeks or months.

surface area How much exposed surface a solid object has.

sustainability Using resources and the environment to meet the needs of people today without damaging Earth or reducing the resources for people in the future.

sustainable Meeting the needs of today without damaging the Earth for future generations.

S-wave A transverse seismic wave through the Earth, produced during an earthquake.

symptom What a person has when they have a particular illness, for example, a rash, high temperature, or sore throat.

synthetic A material made by a chemical process, not naturally occurring.

tectonic plates Giant slabs of rock (about 12, comprising crust and upper mantle) that make up the Earth's outer layer.

telescope An instrument that gathers electromagnetic radiation to form an image or to map data, from astronomical objects such as stars and galaxies. It makes visible things that cannot be seen with the naked eye.

tension A material is in tension when forces are trying to stretch it or pull it apart.

termination When medicine or surgical treatment is used to end a pregnancy.

theory A scientific explanation that is generally accepted by the scientific community.

thermal panel A device that uses the Sun's radiation to heat water.

thermal power station A power station that heats water to produce steam to drive turbines.

tidal power station A power station that uses the tides to drive turbines to generate electricity.

toxic A chemical that may lead to serious health risks, or even death, if breathed in, swallowed, or taken in through the skin.

transformer An electrical device, consisting of two coils of wire wound on an iron core. An alternating current in one coil causes an ever-changing magnetic field that induces an alternating current in the other. Used to 'step' voltage up or down to the level required.

transmitted (transmit) When radiation hits an object, it may go through it. It is said to be transmitted through it. We also say that a radio aerial transmits a signal. In this case, transmits means 'emits' or 'sends out'.

transverse wave A wave in which the particles of the medium vibrate at right angles to the direction in which the wave is travelling. Water waves are an example.

turbine A device that is made to spin by a flow of air, water, or steam. It is used to drive a generator.

ultraviolet radiation (UV) Electromagnetic waves with frequencies higher than those of visible light, beyond the violet end of the visible spectrum.

uncertainty The amount by which a measurement could differ from the stated value.

Universe All things (including the Earth and everything else in space).

unspecialised A cell that has not yet developed into one particular type of cell.

unstable The nucleus in radioactive isotopes is not stable. It is liable to change, emitting one of several types of radiation. If it emits alpha or beta radiation, a new element is formed.

vaccination Introducing to the body a chemical (a vaccine) used to make a person immune to a disease. A vaccine contains weakened or dead microorganisms, or parts of the microorganism, so that the body makes antibodies to the disease without being ill.

variation Differences between living organisms. This could be differences between species. There are also differences between members of a population from the same species.

vein A blood vessel that carries blood towards the heart.

vibrate To move rapidly and repeatedly back and forth.

virus A microorganism that can only live and reproduce inside living cells.

volcano A vent in the Earth's surface that erupts magma, gases, and solids.

voltage The voltage marked on a battery or power supply is a measure of the 'push' it exerts on charges in an electric circuit. The 'voltage' between two points in a circuit means the 'potential difference' between these points.

vulcanisation A process for hardening natural rubber by making cross-links between the polymer molecules.

watt The unit used to measure power: 1 watt = 1 joule/second.

wave speed The speed at which waves move through a medium.

wavelength The distance between one wave crest (or wave trough) and the next.

wet scrubbing A process used to remove pollutants from flue gases.

white blood cell A cell in the blood that fights microorganisms. Some white blood cells digest invading microorganisms. Others produce antibodies.

wind farm A power station that uses the wind to drive turbines to generate electricity.

word equation A summary in words of a chemical reaction.

X-ray Electromagnetic waves with high frequency, well above that of visible light.

XX chromosomes The pair of sex chromosomes found in a human female's body cells.

XY chromosomes The pair of sex chromosomes found in a human male's body cells.

Index

Appendices

Useful relationships, units, and data

Relationships

You will need to be able to carry out calculations using these mathematical relationships.

P1 The Earth in the Universe

distance travelled by a wave = wave speed × time

wave speed = frequency × wavelength

P3 Sustainable energy

energy transferred = power × time

power = voltage × current

Units that might be used in the Science course

length: metres (m), kilometres (km), centimetres (cm), millimetres (mm), micrometres (µm), nanometres (nm)

mass: kilograms (kg), grams (g), milligrams (mg)

time: seconds (s), milliseconds (ms)

temperature: degrees Celsius (°C)

area: cm^2, m^2

volume: cm^3, dm^3, m^3, litres (l), millilitres (ml)

speed: m/s, km/s, km/h

energy: joules (J), kilojoules (kJ), megajoules (MJ), kilowatt-hours (kWh), megawatt-hours (MWh)

power: watts (W), kilowatts (kW), megawatts (MW)

frequency: hertz (Hz), kilohertz (kHz)

information: bytes (B), kilobytes (kB), megabytes (MB)

Prefixes for units

nano	micro	milli	kilo	mega	giga	tera
one thousand millionth	one millionth	one thousandth	× thousand	× million	× thousand million	× million million
0.000000001	0.000001	0.001	1000	1000 000	1000 000 000	1000 000 000 000
10^{-9}	10^{-6}	10^{-3}	$\times 10^3$	$\times 10^6$	$\times 10^9$	$\times 10^{12}$

Useful data

C1 Air quality

Approximate proportions of the main gases in the atmosphere:
78% nitrogen, 21% oxygen, 1% argon

P1 The Earth in the Universe

Speed of light = 300 000 km/s

P2 Radiation and life

Speed of light = 300 000 km/s

electromagnetic spectrum in increasing order of frequency:
radio waves, microwaves, infrared, visible light, ultraviolet, X-rays, gamma rays

P3 Sustainable energy

mains supply voltage: 230 V

Chemical formulae

C1 Air quality

carbon dioxide CO_2

carbon monoxide CO

sulfur dioxide SO_2

nitrogen monoxide NO

nitrogen dioxide NO_2

water H_2O

OXFORD
UNIVERSITY PRESS

Great Clarendon Street, Oxford OX2 6DP

Oxford University Press is a department of the University of Oxford.
It furthers the University's objective of excellence in research,
scholarship, and education by publishing worldwide in

Oxford   New York

Auckland   Cape Town   Dar es Salaam   Hong Kong   Karachi
Kuala Lumpur   Madrid   Melbourne   Mexico City   Nairobi
New Delhi   Shanghai   Taipei   Toronto

With offices in
Argentina   Austria   Brazil   Chile   Czech Republic   France   Greece
Guatemala   Hungary   Italy   Japan   Poland   Portugal   Singapore
South Korea   Switzerland   Thailand   Turkey   Ukraine   Vietnam

Oxford is a registered trade mark of Oxford University Press
in the UK and in certain other countries.

British Library Cataloguing in Publication Data.

Data available.

ISBN 978-0-19-913814-2

10 9 8 7 6 5 4 3 2

Printed by Vivar Printing Sdn Bhd., Malaysia

Paper used in the production of this book is a natural, recyclable product made
from wood grown in sustainable forests. The manufacturing process conforms to
the environmental regulations of the country of origin.

Acknowledgements
The publisher and authors would like to thank the following for their permission
to reproduce photographs and other copyright material:
P13: Chris Schmidt/Istockphoto; **P14**: Alan Schein Photography/Corbis; **P16l**:
Monkey Business Images/Shutterstock; **P16r**: Luca DiCecco/Alamy; **P17**: Richard J.
Green/Science Photo Library; **P18**: Kenneth Sponsler/Shutterstock; **P19**: St. Felix
School, Suffolk; **P20**: source unknown; **P21**: Mehau Kulyk/Science Photo Library;
P22l: Dopamine/Science Photo Library; **P22r**: CNRI/Science Photo Library; **P23**:
Oxford University Press; **P24l**: David Crausby/Alamy; **P24r**: BSIP ASTIER/Science
Photo Library; **P25l**: Dan Sinclair/Zooid Pictures; **P25r**: Zooid Pictures; **P27**: Mauro
Fermariello/Science Photo Library; **P28**: By Ian Miles-Flashpoint Pictures/Alamy; **P29**:
BSIP, LAURENT/Science Photo Library; **P30**: Ariel Skelley/Corbis; **P31**: Mark Thomas/
Science Photo Library; **P32**: Tek Image/Science Photo Library; **P35**: Jim Varney/
Science Photo Library; **P36t**: BSIP, LAURENT H.AMERICAIN/Science Photo Library;
P36b: Pascal Goetgheluck/Science Photo Library; **P37t**: David Scharf/Science Photo
Library; **P37b**: Claude Nuridsany & Marie Perennou/Science Photo Library; **P38**: Dr
Yorgos Nikas/Phototake Inc./Alamy; **P42t**: BSIP, LAURENT/
Science Photo Library; **P42b**: David Scharf/Science Photo Library; **P44**: Kelly
Redinger/Design Pics/Corbis; **P46t**: Charles D. Winters/Science Photo Library; **P46b**:
NASA/Zooid Pictures; **P47**: M.T. Mangan/USGS; **P48l**: David Hardy/Science Photo
Library; **P48r**: Christian Darkin/Science Photo Library; **P49t**: Russell Shively/
Shutterstock; **P49m**: George Steinmetz/Science Photo Library; **P49b**: Dirk Wiersma/
Science Photo Library; **P50t**: John Wilkinson/Ecoscene/Corbis; **P50b**: Harvey Pincis/
Science Photo Library; **P51**: Victor De Schwanberg/Science Photo Library;
P58t: Raoux John/Orlando Sentinel/Sygma/Corbis; **P58m**: Mate 3rd Class Daniel
Scott/U.S. Navy photo; **P58b**: Cordelia Molloy/ Science Photo Library; **P60**: Nick
Hawkes/Ecoscene/Corbis; **P61**: Andrew Lambert Photography/Science Photo Library;
P62t: Burkard Manufacturing Co. Limited; **P62m**: Andy Harmer /Science Photo
Library; **P62b**: Philippe Plailly/Eurelios/Science Photo Library; **P63**: Garo/Phanie/Rex
Features; **P64**: Ian Hooton/Science Photo Library; **P65**: Action Press/Rex Features;

P66: RPL Carburettor and Injection Centre; **P67**: Spencer Grant/Science Photo
Library; **P68**: Simon Fraser/Science Photo Library; **P72**: George Steinmetz/Science
Photo Library; **P74**: Johan Ramberg/Istockphoto; **P78l**: Frank Zullo/Science Photo
Library; **P78r**: Detlev Van Ravenswaay/Science Photo Library; **P79t**: NASA/CXC/
STScI/JPL-Caltech/Science Photo Library; **P79b**: Mark Garlick/Science Photo Library;
P80: Jerry Lodriguss/Science Photo Library; **P81**: Zooid Pictures; **P82**: Chris Butler/
Science Photo Library; **P83t**: Tony Hallas/Science Photo Library; **P83b**: NASA/ESA/
STSCI/R.Williams, Hdf Team/ Science Photo Library; **P85**: Colin Cuthbert/Science
Photo Library; **P86l**: Jack Sullivan/Alamy; **P86m**: Enzo & Paolo Ragazzini/Corbis;
P86r: Sinclair Stammers/Science Photo Library; **P88**: Bettmann/Corbis, **P94**: James
Wardell/Rex Features; **P102l**: NASA/ESA/STSCI/R.Williams, Hdf Team/ Science
Photo Library; **P102r**: AZPworldwide/Shutterstock; **P104**: Bjorn Svensson/Science
Photo Library; **P106t**: Science Photo Library; **P106b**: Guzelian Photographers;
P107: Guzelian Photographers; **P111**: Melanie Friend/Photofusion Picture Library/
Alamy; **P113t**: Getty Images News/Getty Images; **P113b**: Philip Wolmuth/Alamy;
P114: Robert Pickett/Corbis; **P115**: Paul A. Souders/Corbis; **P116t**: Donald R.
Swartz/Shutterstock; **P116b**: Pete Saloutos/Corbis; **P117tl**: Simon Fraser/Mrc Unit,
Newcastle General Hospital/Science Photo Library; **P117tr**: Ed Kashi/Corbis;
P117b: Humphrey Evans/Cordaiy Photo Library Ltd./Corbis; **P118**: Dr P. Marazzi/
Science Photo Library; **P120**: Guzelian Photographers; **P121t**: Science Photo
Library; **P121b**: Biophoto Associates/Science Photo Library; **P122**: AVAVA/
Shutterstock; **P123l**: Bettmann/Corbis; **P123m**: Sipa Press/Rex Features;
P123r: Matt Meadows, Peter Arnold Inc./Science Photo Library; **P125**: Janine
Wiedel Photolibrary/Alamy; **P126**: Dimitri Iundt/TempSport/Corbis; **P127**:
Publiphoto Diffusion/Science Photo Library; **P128**: Martyn F. Chillmaid; **P132**:
Simon Fraser/Mrc Unit,Newcastle General Hospital/Science Photo Library; **P134**:
FotografiaBasica/Istockphoto; **P136**: Chris Hellier/Science Photo Library; **P137**:
Catherine Yeulet/Istockphoto; **P138tl**: Danish Khan/ Istockphoto; **P138tm**:
PeskyMonkey/Istockphoto; **P138tr**: Lee Torrens/Istockphoto; **P138bl**: Yuri Arcurs/
Shutterstock; **P138bm**: Igor Terekhov/Bigstock; **P138br**: Kenneth Sponsler/
Shutterstock; **P139tl**: PhotoCuisine/Corbis; **P139tm**: David Constantine/Science
Photo Library; P**139tr**: Empics; **P139bl**: Taryn Cass/Zooid Pictures; **P139br**: Yves
Forestier/Sygma/Corbis; **P140tl**: David Keith Jones/Images of Africa Photobank/
Alamy; **P140tr**: Dennis Gilbert/VIEW Pictures Ltd/Alamy; **P140bl**: Tom Tracy
Photography/Alamy; **P140br**: Rich Carey/Shutterstock; **P141t**: Masterfile; **P141b**:
Duncan Moody/Istockphoto; **P142t**: Instron® Corporation; **P142b**: J & P Coats Ltd;
P143: Studio 1One/Shutterstock; **P144t**: Tina Chang/Photolibrary; **P144ml-r**:
Andrew Syred/ Science Photo Library, Eye Of Science/ Science Photo Library;
P144b: Eye Of Science/Science Photo Library; **P147**: Science & Society Picture
Library; **P148t**: Dan Sinclair/Zooid Pictures; **P148b**: Taryn Cass/Zooid Pictures;
P149: Tim Pannell/Corbis; **P150l**: Zooid Pictures; **P150r**: Abaca/Empics; **P152**:
W. L. Gore & Associates, Ltd.; **P153t**: Du Pont (UK) Ltd; **P153b**: Eye Of Science/
Science Photo Library; **P154l**: Paul Rapson/Science Photo Library; **P154r**: Paul
Rapson/Science Photo Library; **P156t**: Robert Wisdom/Dreamstime; **P156b**: Eric
Isselée/Fotolia; **P157t**: David Buffington/Photolibrary; **P157b**: Dr P. Marazzi/
Science Photo Library; **P158t**: Irabel8/Shutterstock; **P158bl**: Charles M.
Ommanney/Rex Features; **P158br**: Back Page Images/Rex Features; **P159**: Fact
Fact/Photolibrary; **P162**: PeskyMonkey/Istockphoto; **P164**: Gustoimages/Science
Photo Library; **P166**: Trevor Worden/Photolibrary; **P168**: Terraxplorer/Istockphoto;
P169t: NASA/Science Photo Library; **P169b**: Solent News And Photos/Rex Features;
P170t: David Turnley/Corbis; **P170m**: Silver-john/Shutterstock; **P170b**: CNRI/
Science Photo Library; **P171**: Mike Hill/Alamy; **P172t**: Image Source/Alamy;
P172bl: Astier - Chru Lille/Science Photo Library; **P172br**: Mark Sykes/Science
Photo Library; **P173**: University of Oxford- Division of Public Health and Primary
Health Care; **P175t**: Janine Wiedel/Janine Wiedel Photolibrary/Alamy; **P175b**: Ted
Kinsman/Science Photo Library; **P179t**: Philip Lange/Shutterstock; **P179bl**: Martin
Muránsky/Shutterstock; **P179bm**: The Flight Collection/Alamy; **P179br**:
Photofusion Picture Library/Alamy; **P180t**: British Antarctic Survey/Science Photo
Library; **P180b**: George Steinmetz/Science Photo Library; **P181**: Victor De
Schwanberg/Science Photo Library; **P184l**: David J. Green - lifestyle themes/Alamy;
P184r: lebanmax/Shutterstock; **P186t**: Loskutnikov/Shutterstock; **P186b**: Jerry
Mason/Science Photo Library; **P192**: David J. Green - lifestyle themes/Alamy; **P194**:
Nancy Nehring/Istockphoto; **P196t**: Wayne Bennett/Corbis; **P196m**: Michael
Prince/Corbis; **P196bl**: Kit Houghton/Corbis; **P196bm**: Corbis; **P196br**: Will &
Deni McIntyre/Corbis; **P197t**: EuToch/Corbis; **P197b**: Stephen Ausmus/US
Department Of Agriculture/Science Photo Library; **P198l**: Niall Benvie/Corbis;
P198r: Alex Bartel/Science Photo Library; **P200t**: Alex Segre/Rex Features; **P200b**:
Pakhnyushcha/Shutterstock; **P201l**: HartmutMorgenthal/Shutterstock; **P201r**: Dr
Morley Read/Science Photo Library; **P203**: Wim Van Egmond, Visuals Unlimited/
Science Photo Library; **P204**: Nigel Cattlin/Science Photo Library; **P205t**: Dr Keith
Wheeler/Science Photo Library; **P205m**: Pedro Salaverria/Shutterstock; **P205b**:
Duncan Shaw/Science Photo Library; **P207t**: Oxford University Press; **P207m**: Tom
Brakefield/Corbis; **P207b**: Jeff Lepore/Science Photo Library; **P208t**: Holt Studios
International; **P208b**: Steve Gschmeissner/Science Photo Library; **P216**: Konrad
Wothe/LOOK-foto/Photolibrary; **P217**: source unknown (Georgina Mace); **P218t**:
bl0ndie/Shutterstock; **P218m**: Collpicto/Shutterstock; **P218b**: David Hartley/Rex
Features; **P219t**: Paul Rapson/Science Photo Library; **P219b**: Serhiy Zavalnyuk/
Istockphoto; **P222t**: bl0ndie/Shutterstock; **P222b**: Pakhnyushcha/Shutterstock;
P224: FotografiaBasica/Istockphoto; **P228t**: Kaido Kärner/Shutterstock; **P228b**:
Vincent Lowe/Alamy; **P229tl**: Andrew J. Martinez/Science Photo Library; **P229tr**:
Martin Bond/Photolibrary; **P229bl**: Dirk Wiersma/Science Photo Library; **P229br**:
James King-Holmes/Science Photo Library; **P230t**: Unclesam/Fotolia; **P230bl**:
Pascal Goetgheluck/Science Photo Library; **P230br**: Winsford Rock Salt Mine;
P231: Geographical; **P232**: benicce/Shutterstock; **P234t**: Patrick Frilet/Rex

Features; **P234b**: Timur Kulgarin/Shutterstock; **P235t**: Richard Watson/Photolibrary; **P235b**: Martyn F. Chillmaid/Science Photo Library; **P236**: © Catalyst; **P238t**: Samrat35/Dreamstime; **P238b**: Sean Sprague/Photolibrary; **P239**: American Chemistry Council, Inc.; **P240**: Robert Brook/Science Photo Library; **P242**: Tobias Schwarz/Reuters; **P244**: Kodda/Shutterstock; **P255**: Bob Edwards/Science Photo Library; **P246l**: Craig Holmes Premium/Alamy; **P246m**: Aj Photo/Science Photo Library; **P246r**: aikotel/Shutterstock; **P247**: Peter Ryan/Science Photo Library; **P248t**: Erika Craddock/Science Photo Library; **P248b**: ginosphotos/Istockphoto; **P249t**: Gordon Ball LRPS/Shutterstock; **P249b**: Ton Kinsbergen/Science Photo Library; **P252**: Kaido Kärner/Shutterstock; **P254**: Stephen Strathdee/Istockphoto; **P256t**: Christopher Walker/Shutterstock; **P256b**: ronfromyork/Shutterstock; **P257t**: Tonylady/Shutterstock; **P257b**: Jeffrey Van Daele/Shutterstock; **P258**: Cecile Degremont/Look At Sciences/Science Photo Library; **P259**: Martin Moxter/Photolibrary; **P260**: Ben smith/Shutterstock; **P261t**: Ilya Akinshin/Shutterstock; **P261b**: Foment/Shutterstock; **P262tl**: Rex Features; **P262tr**: Harald Tjøstheim/Dreamstime; **P262ml**: Olexa/Fotolia; **P262mr**: Frances A. Miller/Shutterstock; **P262b**: Jonathan Feinstein/Shutterstock; **P264**: Ieva Geneviciene/Shutterstock; **P265**: Sheila Terry/Science Photo Library; **P267**: Barry Batchelor/PA Photos; **P268t**: Paul Rapson/Alamy; **P268m**: B. S. Merlin/Alamy; **P268b**: © freelights.co.uk 2010; **P269**: © 2009 Dragonfly; **P271l**: Sean Gallup/Getty Images News/Getty Images; **P271m**: Ron Giling/Photolibrary; **P271r**: Mark Sykes/Science Photo Library; **P272**: Steve Allen/Science Photo Library; **P273**: Ria Novosti/Science Photo Library; **P274t**: Marco mayer/Shutterstock; **P274b**: Pearl Bucknall/Robert Harding/Rex Features; **P275t**: D. Kusters/Shutterstock; **P275ml**: hjschneider/Shutterstock; **P275mr**: Rhoberazzi/Istockphoto; **P275b**: Pool/Joao Abreu Miranda/AFP Photo; **P276**: Victor De Schwanberg/Science Photo Library; **P277**: Penimages/Dreamstime; **P279**: Martin Bond/Science Photo Library; **P282**: Htjostheim/Dreamstime.

Illustrations by IFA Design, Plymouth, UK, Clive Goodyer, and Q2A Media.

The publisher and authors are grateful for permission to reprint the following copyright material:

Project Team acknowledgements

These resources have been developed to support teachers and students undertaking the OCR GCSE Science Twenty First Century Science suite of specifications. They have been developed from the 2006 edition of the resources.

We would like to thank David Curnow and Alistair Moore and the examining team at OCR, who produced the specifications for the Twenty First Century Science course.

Authors and editors of the first edition

We thank the authors and editors of the first edition, David Brodie, Jenifer Burden, Peter Campbell, Anne Daniels, Anne Fullick, John Holman, Andrew Hunt, John Lazonby, Jean Martin, Robin Millar, Peter Nicolson, Cliff Porter, David Sang, Charles Tracy, and Jane Wilson.

Many people from schools, colleges, universities, industry, and the professions contributed to the production of the first edition of these resources. We also acknowledge the invaluable contribution of the teachers and students in the Pilot Centres.

The first edition of Twenty First Century Science was developed with support from the Nuffield Foundation, The Salters Institute, and The Wellcome Trust.

A full list of contributors can be found in the Teacher and Technician resources.

The continued development of Twenty First Century Science is made possible by generous support from:

- The Nuffield Foundation
- The Salters' Institute